Googling Earth

Using Google Earth™ to Explore Your World

By JoBea Holt

Cover Image: Google Earth, Image Landsat;
Data SIO, NOAA, U.S. Navy, NGA, GEBCO

Googling Earth: Using Google Earth™ to Explore Your World
Copyright 2012 JoBea Holt

All rights reserved.

googlingearth.com
Published in the United States of America
ISBN 978-1508434573

Updated February 2015

*This is a print-on-demand book on Amazon.
Also available as an eBook.*

Table of Contents

Introduction 5

Finding Me First 6

Setting Up Google Earth 7
   Platform 7
   Internet Connection 7
   Mac vs. PC 7
   Preferences 7
   The View Menu 7
   The Mouse 8
   Help 8

What is Google Earth, Anyway? 9

The Google Earth Tools 13
   Rolling Around Planet Earth 13
   Flying with the Navigation Tools 13
   Finding Yourself (The Statue Bar) 15
   Searching San Francisco 15
   Dropping Down to Street View 17
   Saving My Place 18
   Saving My Place with Pretty Icons 19
   Saving My Place with a Fancy Popup Window 19
   Saving My Life (Folders) 20
   Drawing a Polygon 20
   Creating a Path 21
   Overlaying an Image 21
   Going on a Tour 22
   Traveling Back in Time with Historical Imagery = the Clock 22
   Adding Sunlight 23
   Heading Out of This World 24
   Measuring with the Ruler 24
   Printing Your Favorites 25
   Emailing a Friend 25
   Linking to Google Maps and the Internet Viewer 25
   Saving and Sharing Your Placemarks and Folders 25
   Creating a Reference Scale 26
   Referencing Images 26

What Am I Seeing? 27
   Going on a San Francisco Scavenger Hunt 27
   Finding Patterns in the Land 28
   Diving Into the Ocean 34
   Remembering the Global View 35
   Enjoying a Smoothed Earth 37
   Discovering a Dynamic Earth 37

***My* kmz File** 39

Digging into Layers 40

 Latitude Longitude Grid..40
 Scale Legend and Overview Map...41
 [Borders and Labels], [Places], [Photos] and [Roads]...41
 [3D Buildings]..42
 [Ocean]...43
 [Weather]..43
 [Earth City Lights]...44
 [Africa Megaflyover]...44
 [Rumsey Historical Maps]..45
 [Wikipedia] Backward and Forward..46
 [US National Parks]..47
 Earth Gallery...48
 Flight Simulator..48

Adventures in Google Earth 50
 Exploring Washington DC..50
 Measuring America...52
 Planting Circular Fields..53
 Traveling Through Stories..54
 Mapping Your Past..57
 Taking a Vacation..59
 Navigating the Nile River...60
 Roaming Around Rome...61
 Circling the Globe with Magellan..63
 Climbing High with Pizarro..64
 Paddling Upstream with Lewis and Clark..65
 Bringing World War II to America...67
 Remembering the War in Europe..68
 Finishing the War in the Pacific..69
 Following Current Events...70
 Surviving Extreme Weather...72
 Watching Volcanoes Erupt...73
 Discovering Impact Craters..75
 Digging for Energy..76
 Deforesting Our Planet...77
 Renewing Our Energy...78
 Painting Earth...80
 Investigating Religious Centers...82
 Celebrating Holidays...85
 Playing Sports...87
 Making Movies..91
 Thumbing Through National Geographic Magazine..93
 Launching the Manned Space Program..94
 Flying to the Moon..96
 Landing on Mars...97

Google Earth Books for Teachers..100

Introduction

In 1981, I participated in my first Space Shuttle mission – the second flight of Columbia and the first Shuttle mission to carry a science payload. I was one of the scientists on the mission operations team that monitored and commanded our instrument, an Earth-imaging radar, from a payload control center in NASA's Mission Control Center in Houston. Our interactions with the astronauts in those early days were entirely focused on the engineering success of the mission. By the time our second radar flew in 1984, the astronauts were well trained to photograph the Earth from their unique vantage point, and some of those pictures were used to help analyze the data from our Earth-viewing instrument. By the third and fourth missions in 1994, we trained the astronauts to watch for specific dynamic features on Earth that might help us analyze our data, like snow cover, thunderstorms, and erupting volcanoes. I was one of three Payload Communicators (one person for each of three shifts for 24 hour operations) who talked directly to the astronauts during the mission about their Earth observations.

The excitement of seeing the astronauts' photos of the Earth from space led me to create an educational payload for the Shuttle called KidSat in which a camera dedicated to students was installed in one of the Shuttle's windows. Students in classrooms on the ground controlled the camera, and images were sent electronically to the classrooms during the mission for analysis. These images became the focus of studies in social studies, science, and math, but were limited in coverage to places chosen by a select group of students and one-week missions.

Then along came Google Earth providing a view that only astronauts had in the past – one with the full resolution one would see looking strait down, and yet allowing the sweeping horizon-to-horizon view that provides the context of what you are seeing. Who needs to fly in space anymore?

In 2009 I was asked to write a set of books to help teachers teach their students to use Google Earth, and teach them to see the Earth. I found myself buried in Google Earth for two years exploring history, the settings of children's literature, the paths of explorers, and the impact people have had on our planet. When the books came out (Using Google Earth: Bring the World into Your Classroom, 2012; Levels 1-2, 3-5 and 6-8), I sent copies to all my friends and family expecting them to use them to discover Google Earth, but instead they gave the books to their favorite schools. I realized that books designed for teachers are not ideal for friends, family, and you, so I took what I had learned in writing the teachers' books and wrote this book. I hope you will have as much fun exploring our planet as I had in writing this book.

Finding Me First

The first thing everyone wants to do with Google Earth is find his or her own home. Download Google Earth from http://www.google.com/earth, install it, and open it. Look around the Google Earth Window for the features highlighted in the picture below.

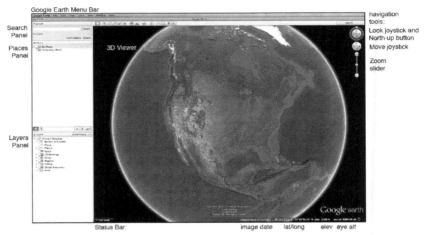

Google Earth Window as a reference for the locations of tools and features.

The part of the Google Earth window that includes the globe and the black background is the "3D Viewer". In the upper left corner of the Google Earth Window, you will find the "Search Panel". Look for the "Search" box and type in your address. Use commas to separate your street address from your city and state or country. Then click the Search button, or just hit return. Google Earth will fly you to your house.

Look in the upper right corner of the 3D Viewer for the "navigation tools". If you don't see them, look in the "Google Earth Menu Bar" under "View" → "Show Navigation", and choose "Always" or "Automatically". Click on the "Zoom slider" (the bar with the + and – on it) to zoom in and out of your yard. Cool, huh?

Find a few more familiar places by simply typing their famous names into the Search box in the Search Panel. Sometimes you may need to zoom in or out to see the structure.

The Statue of Liberty, New York
Eiffel Tower, Paris, France
Sydney Opera House, Sydney, Australia
Disneyland, California
Cape Hatteras, North Carolina (the picture on the cover of this book)

If you thought that was fun, it is only the beginning!

Setting Up Google Earth

Here are a few more tips to help you get set up to explore with Google Earth.

Platform

Google Earth can be run on a desktop or laptop computer; or on a notebook like an iPad. This book addresses the computer version of Google Earth. The iPad version was quite different at the time this book was written, but much of what is included here is applicable. Look for *Googling Earth with Your iPad* - coming soon!

The specific system and Internet requirements for PC, Mac and Linex systems are outlined by Google at http://www.google.com/earth/download/ge/. If Google Earth is not working on your computer, it is a good idea to look at the requirements.

Internet Connection

To accommodate the streaming data that make Google Earth so engaging, make sure you have a reliable high-speed Internet connection.

Mac vs. PC

In general, the interfaces for Macs and PCs are the same for Google Earth. There are some small differences that are listed below. This book was written for Macs, so if you are a PC user, watch for these very small differences.

- The small arrows to the left of folders and layers on Macs that allow you to open and close them are plus signs (+) on PCs.

- When you right-click on a Placemark to edit it, the popup window will list "Get Info" for a Mac, and "Properties" for a PC.

Preferences

Look through the Preferences in the Google Earth Menu Bar under Google Earth → Preferences to select your favorite measurement units, but for the rest of the parameters, start with the default settings until you become more familiar with how Google Earth operates.

The View Menu

Look along the top of the Google Earth Window in the "Google Earth Menu Bar" for "View". A variety of settings are available and most will be investigated in this book. To begin, check "Toolbar", "Sidebar", "Status Bar", "Atmosphere", and "Water Surface".

The Mouse
You will often use the right-click option on your mouse. If you have a track-pad or a single-click mouse, you can use control-click (hold down the "control" key on your keyboard and click the track pad or mouse) instead.

Help
Help and more information about how to use Google Earth are available on http://support.google.com/earth.

What is Google Earth, Anyway?

Beginning with the tiny windows in the Mercury capsules, astronauts and cosmonauts have been fascinated with their unique view of Earth. Many spend their free time floating over the windows of the Space Shuttle or the International Space Station watching the world go by, and some of the most precious treasures from any mission are astronauts' photographs of Earth, just as your most precious treasures from a vacation are your pictures. My favorite book of astronaut photos is Orbit by Jay Apt (2003), and my favorite website is http://eol.jsc.nasa.gov/.

Tracy Caldwell Dyson of the Expedition 24 crew looks out the Cupola window in the International Space Station. (Photograph courtesy of NASA, http://apod.nasa.gov/apod/ap101115.html)

Astronaut Kathy Sullivan remembers,

" *Both the sweep of the view and the detail you can see are quite remarkable...In even a brief period at the window, you can get quite an introduction to all the diverse, interrelated Earth science disciplines - geography, geology, oceanography, ecology, meteorology - as this great planet slides before your eyes, revealing an ever-changing array of forms, patterns, and colors.*" (http://eol.jsc.nasa.gov/newsletter/uft/uft2.htm)

In 1960, Earth-imaging satellites were launched to monitor the weather, and in 1972, the first Landsat satellite was put in orbit to photograph the Earth for land-use applications. The earliest satellites carried cameras that operated in the visible part of the electromagnetic spectrum; current missions carry instruments that operate in the visible, infrared, and microwave regions. Early analysis of the images involved a person looking at the pictures and identifying unique features. Current data analysis deals with millions of times the data of the earlier missions and uses complex programs to automatically map and monitor specific features like sea surface temperature, snow cover, and atmospheric water vapor. To learn more about the National Aeronautics and Space Administration (NASA) satellites that are monitoring our planet, go to http://earthobservatory.nasa.gov/GlobalMaps/ and http://climate.nasa.gov/Eyes/.

But with all the fancy instruments in space, people remain fascinated with looking at color pictures of Earth and love to discover features with which they are familiar. So Google Earth was born. Google Earth presents images of Earth at a wide range of resolutions from the full globe to the scale of cars and animals. To do this required collecting a high-resolution picture of Earth. Satellites typically can image only about a 100 x 100 mile square region at a time, so these regional images had to be mosaicked together as you might put together a puzzle. Google first mosaicked a lower resolution data set and then added high-resolution images as they were acquired. Often regions were covered with clouds, so in some cases, it took a long time to get a cloud free image for the puzzle. These early data sets were provided by the Defense Meteorological Satellite Program (DMSP), NASA, and the National Oceanic and Atmospheric Administration (NOAA), and had resolutions of no better that 30 m.

But people wanted to see more detail; 1 to 5 m resolution imagery was desired. In 1997, a company named Digital Globe launched a series of commercial satellites designed specifically to collect very high-resolution images globally. These satellites were called Early Bird 1 (3 m resolution), QuickBird (2.4 m resolution), and WorldView-1 and -2 (.5 m resolution). Much of the Google Earth globe uses this high-resolution imagery, as well as the French Space Agency's (Centre National d'Etudes Spatiales, CNES) 2.5 m SPOT images.

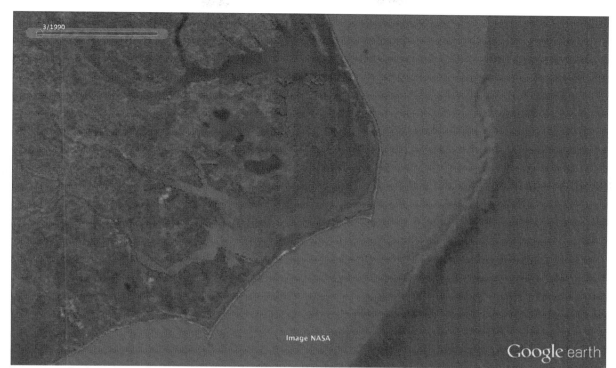
Low resolution Google Earth view from a Landsat image of Cape Hatteras from the 1990s. (Google Earth Image NASA, 35°22'25"N 75°54'34"W, eye alt 180 mi)

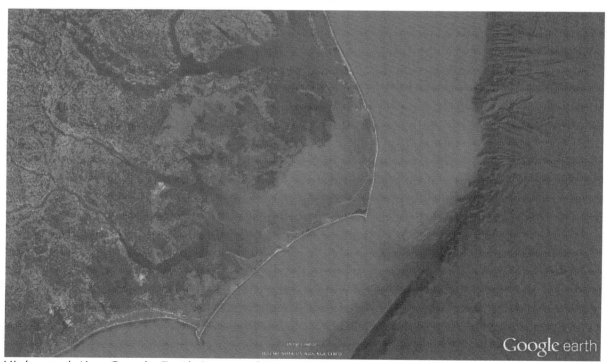
High-resolution Google Earth image of Cape Hatteras from 2012. (Google Earth Image 2012 TerraMetrics, Data SIO, NOAA, U.S. Navy, NGA, GEBCO, 35°22'26"N 75°54'34"W, eye alt 180 mi)

As the satellites continued to fly and collect more and more data over each place on Earth, a historical imagery feature was added to Google Earth, which allows users to look at multiple images of a specific region collected on different dates. In some regions, aerial photography has also been added to provide views that date back to the 1930s.

Three-dimensional view of Mount Rainier, Washington using topographic data that is now part of Google Earth. (Google Earth Image USDA Farm Service Agency, 2012 TerraMetrics, 2012 Cnes/Spot Image, 46°48'02"N 121°46'26"W, eye alt 20000 ft)

In 2000, a Space Shuttle mission carrying an instrument called the Shuttle Radar Topography Mission (SRTM) collected a digital topographic data set of the entire land surface between 60°N and 60°S latitudes. This data set was overlaid on Google Earth to provide three-dimensional (3D) views of the Earth's surface.

With the global high-resolution 3D date set in hand, viewing it became the challenge. The Google Earth software tool was created that allows a user to navigate around the globe, zoom in and out to see the surface at a wide range of resolutions and perspectives, and click back in time to see how the surface has changed over years and decades. Abstract data layers have been added to display latitude and longitude, day and night, and roads and city names, among other features. Photographs, three-dimensional models of structures, and YouTube videos collected and created by people all around the world enhance the global map with a more familiar local view. Tools that allow a user to save places they have explored and share their discoveries with friends complete the features that make Google Earth awesome.

The Google Earth Tools

You are now in the same category as most people I know who have tried Google Earth: you have found your house and a few other famous places. But there is so much more to explore. Follow these introductions to the Google Earth tools using San Francisco as a reference. In addition to learning about the tool, you will discover something special about the City by the Bay.

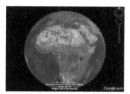

Rolling Around Planet Earth

If your house or another famous place still fills the 3D Viewer, use the Zoom slider to zoom back out to a global view. To begin investigating Earth, use your mouse to move the cursor in the 3D Viewer to roll the globe (click down on the mouse and slide the cursor across the globe). Look for the continents, the colors of the major deserts (the Sahara, the Arabian Desert, and the red deserts of Australia), the forests (the Congo, the Amazon, and the boreal forests of Canada and Russia), the ice covered regions (Greenland, Antarctica, and the peaks of the Himalayas), and Earth's thin atmosphere on the horizon. Notice how large the Pacific Ocean is and think about how long it took Magellan to cross it. Notice how the deserts and forests of Africa change with latitude. See if you can find San Francisco from this global perspective. Think about regions you would like to see in more detail – but, as you travel through this book, never forget to return to the global view.

Flying with the Navigation Tools

Look in the upper right of the 3D Viewer for the navigation tools. You have already used the Zoom slider. Use it along with your cursor to zoom in to get a closer look at San Francisco, or search for San Francisco using the Search Panel. Look for San Francisco Bay, the peninsula on which the city sits, Treasure Island, the bridges, and Golden Gate Park (hint: look for a large green rectangular area).

San Francisco showing Treasure Island, San Francisco Bay, the bridges, and Golden Gate Park. (Google Earth Image 2012 TerraMetrics, CSUMB SFML, CA OPC, SIO, NOAA, U.S. Navy, NGA, GEBCO, 37°47'34"N 122°25'13"W, 24 mi)

The joystick in the navigation tools with the eye on it is called the "Look joystick". Zoom out to a global view and notice what happens to Earth when each of the small arrows on the Look joystick is clicked. Use your cursor to rotate the outer ring on the Look joystick (the "North-up button") to change the cardinal orientation of the globe. Click once on the N on the North-up button to return to a north-up view, which is usually more familiar.

To investigate the three-dimensional aspects of Google Earth using the Look joystick, zoom back in to the southeast corner of Golden Gate Park until the rust-colored football field fills the 3D Viewer. Click on the top arrow on the Look joystick until you can see the horizon and note how your perspective changes. To look around from your new perspective, rotate the North-up button. Do you see the hills of San Francisco? To return to a view from above, click on the bottom arrow on the Look joystick, or on the "u" key on your keyboard.

The joystick with the hand on it is called the "Move joystick". Zoom out to a global view and click on each of the four little arrows on this joystick to discover which way the Earth rotates. Zoom back in to San Francisco and use the Move joystick to travel north, south, east and west. Hold the cursor down and move it around the edge of the Move joystick to fly around the San Francisco area. Notice how the hills look three-dimensional as you fly over them.

Try out your new tools by using the Search Panel to search for Lone Pine, California, just over the Sierras from San Francisco. Lone Pine is in the Owens Valley, which lies between the two mountain ranges. The Sierras are to the west and the White Mountains are to the east. Zoom in further until one city block in Lone Pine fills your 3D Viewer. Click on the top arrow on the Look joystick for a new view of the mountains surrounding Lone Pine. Use the North-up button on the Look joystick to look around Owens Valley, and use the Move joystick to move closer to the Sierras. Can you find Mt. Whitney, the tallest mountain in the lower 48 states? Return to a view from above by pressing the "u" on your keyboard.

`Imagery Date: 4/10/2013   42°12'18.72" N  100°39'26.64" W  elev  3046 ft   eye alt 1421.25 mi`

Finding Yourself (The Statue Bar)

Look along the bottom of the 3D Viewer for the "Status Bar", which provides information about the location of the image and your eyes. First find latitude and longitude ("lat/long"; e.g., 42°12'18.72"N 100°39'26.64"W). If your cursor is in the 3D Viewer, this is the lat/long of the cursor. If you move your cursor out of the 3D Viewer, this is the lat/long of the center of the 3D Viewer.

Elevation, or "elev", is also shown in the Status Bar. This is the elevation of the land relative to sea level either at the location of the cursor if the cursor is in the 3D Viewer, or of the center of the 3D Viewer. The "eye alt" at the right of the Status Bar shows the altitude above sea level of your eye if you were viewing Earth from a plane or a satellite. Search for Death Valley and then Mount Everest and use the navigation tools to move around. Note how eye alt and elevation change as you move.

Finally, Image Date is included in the Status Bar. This is the date the image was acquired.

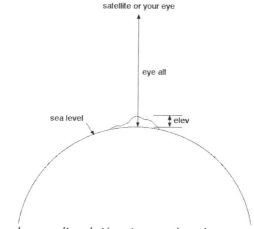

Drawing showing elevation and eye alt relative to sea level.

Searching San Francisco

You have already used the Search Panel to find your house and a few other famous places. You can investigate this tool further using the Grateful Dead

House in the famous Haight Ashbury district of San Francisco, where hippies once congregated. The address of the Grateful Dead House is 710 Ashbury St., San Francisco, California. Type it in the Search box and use the navigation tools to look around. Note that it is always a good idea to include the city and state, or city and country in your search to make sure you end up in the right place.

You may want to visit the Grateful Dead House and need directions. Look in the Search Panel for the Get Directions button and click it. You will now see two boxes – one for your starting location (A) and one for your destination (B). Type the addresses into the boxes. Then click the Get Directions bar, or hit return. You will see a line in the 3D Viewer showing the roads you would take to travel by car from your house to the Grateful Dead House, and street-by-street directions in the Search Panel. At the bottom of the directions is a link to a Printable view. Sometimes, at the top of the directions, you will see four buttons that allow you to find your directions using a car, public transportation, your feet or your bicycle. Click on the "X" when you are done finding your directions.

Notice also the "History" button next to the Directions button. If you click this button, a list of all the places you have searched for in your current session will be displayed in the Search panel.

Sometimes when you type in a location or a city, Google Earth has a hard time determining the place you really want to see. If you want to find the Great Pyramids of Giza in Egypt, for example, and type in "Pyramid", Google Earth will show you a view that includes the Great Pyramids along with several other locations that are named after the famous pyramids. Try it and look for the list of these places below the Search box in the Search Panel, and the red balloons related to each location in the 3D Viewer. You may be able to decide which location you really want by examining the balloons, or you may need to look through the list of names. Once you find the correct one, double click on it in the Search Panel and you will be taken there.

You can also search by latitude and longitude. To find California, you only need to use degrees. Type 37N 122W (or 37°N 122°W) in the Search box and hit return. You will need to zoom out to see all of California. To find San Francisco, you will need degrees and minutes. You can avoid typing in the degree and minute symbols in the Search box; use blank spaces instead: 37 44N 122 27W (or 37°44'N 122°27'W). Zoom until the city fills your screen. To find the Golden Gate Bridge, you will need to include degrees, minutes and seconds: 37 49 04N 122 28 42W (or 37°49'04"N 122°28'42"W), and to find the tollbooth for the bridge, you will need to include decimal seconds: 37 48 26.04N 122 28 32.27W (or 37°48'26.04"N 122°28'32.27"W). This time you will need to zoom in.

Dropping Down to Street View

Perhaps you noticed a little orange figure above the Zoom slider in the navigation tools. The figure is called the "Pegman" because it looks like a clothes peg. When the Pegman appears, views from the street level are available. Navigate to a view from above the Golden Gate Bridge. Use your cursor to drag the Pegman into the 3D Viewer and notice that some streets turn blue. These are streets that have been photographed by a special panorama camera on a Google Earth Car. If you drag and drop the Pegman on one of the blue streets, or onto the Golden Gate Bridge since you are there anyway, you enter "Street View", which means you are looking at images taken by the car camera, not satellite images. Use the Move joystick and the North-up button to move and look around. Or just click on a location in the image and you will move to that location. To exit Street View, click the button in the upper right of the 3D Viewer. Return to the Grateful Dead House and investigate the restored Victorian architecture of the houses on Ashbury Street using Street View.

Note that in some cases, Street View offers two options: Street View and "Ground-level View". You can switch between the two using the button in the upper right of the screen once you enter Street View. Ground-level view uses satellite imagery, and Street View uses photography collected using the panorama camera that travels on the Google Earth Car. Later when you discover Layers, click on 3D Buildings for a more interesting Ground-level View.

An entertaining place to use Street View is Sea World in Orlando, Florida. After searching for Sea World, Orlando, Florida, look for Shamu's tank (the largest blue pool with an arc-shaped white roof over the grandstand) and drop the Pegman on the blue line over the roof of the grandstand. Use the North-up button and the Look joystick to find the whales. Also look at Street View in some of California's National Parks including Yosemite, Redwood, Sequoia, and Death Valley; Pompeii, Italy; Lago Tumbira, Amazones, Brazil (taken from a boat!); and Half Moon Island, Antarctica.

Google Earth is now using a Google Earth Trike to obtain street view images of remote regions like the Arctic. Search for Deadhorse Airport, Prudhoe Bay, Alaska to find one such region photographed using the bicycle. Watch for new ways to explore using Street View as a Google Earth Underwater Camera begins collecting imagery.

The Google Earth car and camera used to take Street View photographs. (http://en.wikipedia.org/wiki/File:GoogleStreetViewCar_Subaru_Impreza_at_Google_Campus.JPG)

Saving My Place

You have found your home, a few special places, and some San Francisco landmarks. Perhaps you have already forgotten where you have been. You can save the places you discover and find them easily the next time you look at Google Earth on your computer using the "Places Panel". Begin by using the Search Panel to find AT&T Park, San Francisco, California. Look in the "Toolbar" just above the 3D Viewer for the little yellow thumbtack. This is the "Placemark" tool. Click on it and a Placemark window and a Placemark icon (yellow thumbtack) will appear on your screen. Use your cursor to move the yellow thumbtack to the exact place you want to mark. Type "AT&T Park, San Francisco, California" in the "Name:" box in the Placemark window. Click OK in the lower right of the Placemark window and look in the Places or Search Panel for your Placemark. If it is in the Places Panel, leave it there. If it is in the Search Panel, use your cursor to drag it to the Places Panel. Note that you can open and close the Search and Places Panels using the little arrows to the left of the panel names. Roll the Earth to another continent, and then click (or double click) on the AT&T Park Placemark in the Places Panel and you will be flown back to the ballgame.

Let's pretend that you want your Placemark to show a closer view of the ballpark. To edit your Placemark, use your mouse to right-click (or control-click) on it in the Places Panel and choose Get Info from the pop-up window (or "Properties" if you are using a PC). Use the navigation tools to find a closer view, and then look in the Placemark window for the View tab and click on it. Click "Snapshot current view", and then click OK to close your Placemark. Again, roll the Earth to another continent, and click on the AT&T Park

Placemark in the Places Panel and you will be flown back to the new view of the field.

Try creating Placemarks for your home and some of the famous places you have visited. Search for your old schools, previous homes, or your parents' or children's homes, and placemark them also. Collect all your Placemarks in the Places Panel. Close Google Earth and open it up again to see that your Placemarks are still there.

Saving My Place with Pretty Icons

In addition to the yellow thumbtack, a variety of icons in a variety of colors and sizes are available for making Placemarks. After opening the Placemark window, look in the upper right corner of the window for the yellow thumbtack and click on it. A new window of Placemark icons will appear. Choose one and select a custom color and size using the drop-down options at the top of the Placemark Icon window. Then select OK and continue creating your Placemark.

Note that you can also create a Placemark with no icon by clicking on No Icon at the bottom of the Placemark Icon window, or you can create an icon with no name by leaving the Name: box in the Placemark window blank.

Saving My Place with a Fancy Popup Window

You can add websites, pictures, and text to the Description in the Placemark window, and then the Description will show up in a pop-up window when you click on that Placemark. Be aware that lots of pictures can slow down Google Earth. Also note that the first line or two of the text in the Description shows up under the Placemark in the Places Panel and can make your Places Panel more chaotic.

You will need to use a few html tricks to make your Description readable:

To skip to the next line: add a
To skip to the next paragraph: add a </p>
To add a picture that is stored on your computer:
 for PC
PC example:
<img src="C:\Documents and Settings\HP\MyDocuments\Pictures\image.jpg">
<img src=*name of image*> for Mac
Mac example:

<img src=image.jpg>
To add a picture from the Internet: find the image, copy the url, and, in the Placemark window, click on the "Add image..." button.
To add a link to the Internet: find the link, copy the url, and, in the Placemark window, click on the "Add link..." button.

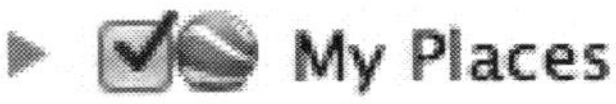

Saving My Life (Folders)

You have likely gathered quite a few Placemarks in your Places Panel. These can be organized into folders and subfolders. To begin, look at the top of the Places Panel for the [My Places] folder and right-click on it. Choose Add and then Folder (Add → Folder). A Folder window will appear. You can name it "Me", or use your name, to collect all the Placemarks about yourself. Close the folder by clicking OK and look for it in the Places Panel. Use the cursor to drag the Placemarks related to you into the [Me] folder. You can expand or hide the contents of a folder using the little arrow (or the (+) if you have a PC) to the left of the folder in the Places Panel, and you can show or hide the Placemarks in the 3D Viewer by checking on the little box to the left of the folder.

To add a second folder, right-click again on the [My Places] folder and choose Add → Folder. Name this folder [Famous Places], and then drag the famous places you have visited on Google Earth into this folder. As you collect new Placemarks, you may want to create additional folders and subfolders focused on your personal interests, or you may want to change the folders you created here to entirely different folders.

Perhaps you want to separate Placemarks related to you from those related to your relatives. Right-click on the [Me] folder and choose Add → Folder. Name this folder [My Life]. Right-click again on your [Me] folder and choose Add → Folder. This time name the folder [My Family]. Drag the Placemarks in your [Me] folder into their appropriate subfolder. Additional third and fourth tier folders may be made as your collection of Placemarks grows.

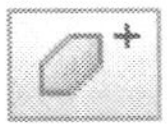

Drawing a Polygon

In addition to marking a location with a Placemark, you can also mark it with a polygon. To illustrate, you can make a polygon around Golden Gate Park. Search for the park and center it in your 3D Viewer. Look in the Toolbar for the "Polygon" tool and click on it. The Polygon window is very similar to the Placemark window. Use your cursor to mark the corners of your polygon – it can have anywhere from three to thousands of sides. As you mark the polygon, a white shape appears in the 3D Viewer. To make your shape a different color, less opaque or only an outline, look under the Style, Color tab in the Polygon window and click your selections. To make the polygon an outline only, choose 0% for "Area" opacity. Click OK when your polygon is complete. Note that the

Polygon appears like a Placemark in the Places Panel but has a little polygon shape next to it rather than a yellow thumbtack.

Creating a Path

Beyond Placemarks and Polygons, you can also keep track of a location by making a path. This is particularly useful if you are tracking the path of an explorer, the plot of a book, or your summer vacation. Zoom to a view that includes the Bay Bridge and the Golden Gate Bridge (eye alt about 10 mi). Navigate to the Oakland side of the Bay Bridge (the east end), click on the "Path" tool in the Toolbar, and click once to start your Path. Click along the bridge to continue your Path, turning at Treasure Island, and continue across until you get to San Francisco. Click westward along the shore until you get to the Golden Gate Bridge and then click across the Golden Gate Bridge. Note that if you want to change the view in the 3D Viewer while you are making a Path, you need to use the navigation tools, *not* the cursor. If you accidentally make an unwanted point, right click (or control-click) on it. Name your Path anything you want and click OK. Look for your Path in the Places Panel; it will have a Path symbol next to it.

Right-click on the Path in the Places Panel and choose Get Info to open the Path window again. Click on the Measurement tab to see how far you would have to walk to cross both bridges. You can change the color and thickness of your Path line using the Style, Color tab. If you want to add to your Path, move your cursor over the last point until it changes color, and begin adding points. If you want to move a point, move your cursor over that point until it changes color and drag it to a new location. If you want to add points between two existing points, click on the first of the points and then click to add your new points.

Overlaying an Image

To illustrate the "Image Overlay" tool, go to the National Park website for Alcatraz Island and look for the map under Plan Your Visit (http://www.nps.gov/alca/planyourvisit/maps.htm). Click on "Open the PDF of the island map by clicking here" and do a screen capture of the map. Save the screen capture on your desktop. Open Google Earth and search for Alcatraz Island, California. Zoom to an eye alt for which the island fills your screen. Choose the Overlay tool in the Toolbar and use the popup window to Browse for your Alcatraz screen capture on your desktop. Select it and you will see the map on top of the Google Earth image of the island. Look in the Overlay window for a slider – you can slide it to change the opacity of your overlay map, which makes it easier to match up with the Google Earth view. You can grab and move the corners of the map with the cursor to "co-register", or match, it to the Google Earth view. You can also use the little green diamond to rotate the map. When you are happy with the match, slide the slider back to an opaque view and click "OK". Look for your Overlay in the Places or Search Panel. Click it on and off to help you identify the buildings on Alcatraz that the prisoners knew so well.

Going on a Tour

If you have already created a Path, you can tour it by highlighting the Path in the Places Panel and clicking on the "Play Tour" button at the bottom right of the Places Panel (it looks like the Path tool).

If you want to make a tour over which you have more control, you need to start with a plan. The best way to is make a folder of the Placemarks you want to tour and then organize them in the order you want to view them. As an example, make a folder with Alcatraz, Golden Gate Park and AT&T Park. Make sure each Placemark has a view with the appropriate eye alt and perspective to make your tour interesting. Click on the "Tour" tool in the Toolbar and look in the lower left corner of the 3D Viewer for the Tour controls. Double click on your first Placemark, which will be the start of your Tour. Click the red button on the Tour controls to start recording, and stay at this Placemark for a few seconds. Then double click on your next Placemark; Google Earth will fly you to that Placemark and record the flight. Stay at that Placemark for as long as you like and then proceed to the next Placemark. If you want to navigate around your Placemark as part of your tour, use the navigation tools. When you have seen your last Placemark, click the red Play Tour button again to stop your Tour.

A new "Tour Playback" window will appear in the lower left of the 3D Viewer and your Tour will begin to replay automatically. Click the button on the far right of this window to save your Tour; a popup window similar to the Placemark window will appear so you can name your Tour. Look for your Tour in the Places Panel and move it to the appropriate folder.

Traveling Back in Time with Historical Imagery = the Clock

The "Historical Imagery" tool, or "Clock", is one of the coolest features of Google Earth. It will lead you to imagery taken in the past, sometimes as far back as the 1930s (using aircraft photos, not satellite images). Search for Candlestick Park in San Francisco and zoom to an eye alt of about 5,000 ft. Click on the Clock tool and look for the Clock slider in the upper left of the 3D Viewer. Move the slider back in time to 1938 to see where Candlestick Park was built. Move the slider forward to 1993 when the park was used by both the San Francisco Giants and the San Francisco 49ers. Slide ahead to the current time to see what the field looks like as a dedicated football stadium, as the Giants moved to AT&T Park in 2000.

Note that the Clock slider indicates the latest date that imagery in the 3D Viewer could have been taken. The Imagery Date in the Status Bar indicates the earliest date the imagery could have been taken. So the imagery in the 3D Viewer would have been taken sometime between the two dates.

Other dynamic places to follow with the Clock are:
Yankee Stadium, New York City, New York – *Look for both the new and the old stadiums.*
Birds Nest Stadium, Beijing, China – *Built for the 2008 Olympics.*
The World Trade Center, New York City, New York – *Destroyed on September 11, 2001.*
Warsaw, Poland (52°14'09.34"N 21°02'26.81"E) – *The bridge was bombed during WWII and bomb craters can be seen in the area at 52°14'20.69"N 21°02'42.90"E, which is now a stadium.*
Palm Island, Dubai, United Arab Emirates – *After watching this strange-looking island take shape, look around Dubai for other new constructions, including The World Islands, Ferrari World, and the Burj Khalifa.*
Las Vegas, Nevada – *Watch the entire city grow over time, as well as the Strip.*
The Aral Sea (eye alt 250 mi) – *Even regions change over time. The water that feeds this lake has been diverted for farmland over the years.*

Adding Sunlight
When you look at the Google Earth globe, it is always in full sunlight, which, of course, is not the way the world works. You can add the dark side of Earth, or day and night, using the "Sunlight" tool in the Toolbar; click on it and watch your view of Earth change. If the globe still seems to be completely lit, use the cursor to roll it around because you may be looking directly at the sunlit side of the Earth. Look for your city and compare the shadow on the globe to where the sun is outside your window.

When you open the Sunlight tool, the shadow on Earth is the current shadow based on how far Earth has rotated relative to the Sun today. You can adjust the time of day using the Sunlight slider in the upper left corner of the 3D Viewer. As you move the slider and watch the shadow change, remember that, in reality, Earth rotates relative a fixed Sun, while in the 3D Viewer, it appears as if the Sun is revolving around Earth. Search for the North Pole and slide the Sun through a day. Think about the current season and whether the pole should be in complete darkness (northern winter) or complete daylight (northern summer). Search for the South Pole and compare the lighting to the North Pole. This comparison is most interesting in winter and summer, and less interesting in the spring and fall.

The range of time covered by the Sunlight slider can be changed using the tiny magnifying glasses with the + and - signs. Click once on the - magnifying glass to change the range to a week, again to see a month, and once more to see a year. Slide the slider for a year from January to December and watch how the sun at the North Pole changes. To watch this without sliding the slider, click on the tiny clock on the Sunlight slider. I know of no better way to explain the seasons than to use Google Earth's Sunlight tool!

Heading Out of This World

In addition to imaging most of Earth in high resolution, NASA has collected extensive imagery of the Moon and Mars in high resolution. Using the same approach of mosaicking images onto a globe, Google Earth allows you to fly round the Moon and Mars. Click on the "Planet" tool in the Toolbar and choose Moon. You will be taken to a global view of the near side of the Moon. Roll the Moon to a center lat/long of 0 0, and zoom in until the Moon fills the screen and North is up. Look for the dark and light patterns that create the face of the Man in the Moon. These patterns are created by solidified lava plains that were once thought to be oceans, and light-colored highlands. Navigate to the opposite side of the Moon and compare the number of craters on the near and far sides.

Return to the Planet tool and choose Mars and you will be taken to the red planet. Zoom in and navigate around – does Mars have as many craters at the Moon?

You can also navigate around the whole sky using the "Sky" option in the Planet tool. My favorite layer for the Sky is the "Our Solar System" in which you can see each planet.

Later you will learn more about the Moon and Mars using Google Earth.

Measuring with the Ruler

The Ruler tool can be used to measure anything from the length of a car to the circumference of Earth. Return to Earth and search for the "San Francisco Airport". Zoom in until the runways fill your screen and North is up. Click on the Ruler tool in the Toolbar and select units of miles. To measure the length of a runway, use your cursor to click on one end of a runway. Watch the numbers change as you move the cursor to the other end of the runway. Click on the opposite end of the runway to see its full length.

The Ruler tool can also be used to measure heading, or the angular direction, relative to north. Click in the middle of one of the runways and watch the heading as you measure it in the westward direction. How does the heading change if you measure it in the eastward direction?

Now try a global view. Use the Ruler to measure the distance from one horizon to the opposite horizon. Are you measuring Earth's diameter or half of Earth's circumference?

Take a look at the Ruler units once more. Have you ever heard of the units of "smoot"? Look it up on Wikipedia!

Printing Your Favorites
You can print your current view in the 3D Viewer or descriptions of the Placemarks in a selected folder using the "Print" tool in the Toolbar. Remember that printing high-resolution images of Earth requires a good printer and photo quality paper, and uses a lot of ink!

Emailing a Friend
A view in the 3D Viewer, a Placemark, or a folder can be emailed to a friend by clicking on the "Email" tool in the Toolbar. The Email popup asks you to choose to send a picture, a link to your current view, or, if you have already highlighted a Placemark or folder, that Placemark or folder. Try sending each option to yourself to decide which option you would prefer the next time you get excited about something you have found on Google Earth and want to send it to a friend.

Linking to Google Maps and the Internet Viewer
You can link directly to Google Maps using the "Google Maps" tool in the Toolbar. Your 3D Viewer will change to an Internet Viewer. The view in Google Maps will match your current view in the 3D Viewer. To return to the 3D Viewer, click "<<Back to Google Earth" in the upper left of the Internet Viewer.

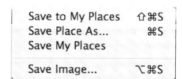

Saving and Sharing Your Placemarks and Folders
When you add a Placemark (or Path, Overlay, or Tour), it is automatically saved on the computer you are currently using as long as it is in the Places Panel. You can also save a Placemark, or a folder of Placemarks, in any location on your computer. The Placemark can then be moved to a different computer or emailed to a friend. To do this, first highlight a folder or Placemark you want to save, and then look in the Google Earth Menu Bar under File for "Save", and choose "Save Place As..." A popup window will appear where you can choose where to save the Placemark or folder and what to name it. The file will be a "kmz" file – kmz is a zipped version of a kml file; kml stands for keyhole

markup language, the computer language of Google Earth. If you are saving the folder as a backup, you might want to add a date to your file name.

If you have not highlighted a Placemark or folder, you can still save the image in the 3D Viewer. Under File → Save, choose "Save image..." and you will be able to save the image as a jpg file.

Try emailing a file to yourself on another computer. When you get the email, click on the kmz file in the body of the email and it should open up in Google Earth. Your new folder will be in the [Temporary Places] folder in the Places Panel and will disappear unless you move it into the [My Places] folder. Be sure you have Google Earth on the second computer!

Creating a Reference Scale

As you zoom in and out of Google Earth, it is easy to lose track of the size of features relative to things with which you are familiar. Consider having a few references in mind, like the size of your house or city block in feet, or one mile. Zoom to your house and then your city block. Use the Ruler to measure one dimension and then round it to an easy number to remember. Then when you search for a new location, like The Sydney Opera House or the Grand Canyon, you will be able to compare their size to something familiar.

Referencing Images

Acknowledgement for the source of each image is presented in the lower center of the 3D Viewer. If more than one image makes up the view in the 3D Viewer, all references are included. If you want to use a Google Earth image, be sure to read the permission guidelines at http://www.google.com/permissions/geoguidelines.html and acknowledge use of the image using these references.

What Am I Seeing?

Now that you know how to use the basic Google Earth tools, let's make sure you know what you are seeing. I was once on an airplane and a boy who was about ten years old was looking out the window as we flew over the Mojave Desert. All of a sudden he declared with great satisfaction, "Look, Mom, the Rocky Mountains!" We were still over the Mojave. I do recall on my first plane flight when I was six being amazed at how much snow there was. I was actually looking at clouds. So what are you really looking at in Google Earth?

USS Hornet, Oakland, California (Google Earth Image 2012 Google, 37°46'24"N 122°18'11"W, eye alt 2000 ft)

Going on a San Francisco Scavenger Hunt

Within a populated area, there are manmade features that are easy to identify wherever you are. Below is a list of some of these in the San Francisco area. You can find them by using the Search Panel. Most are best viewed at an eye alt of 2000 ft or less, unless otherwise noted.

Small park (Alamo Square Park)
Football field (37°44'45"N 122°28'48"W)
Tennis court (Golden Gate Park Tennis Courts)
Golf course (Presidio Golf Course, eye alt 5000 ft)
Baseball field (Ulrich Field and Benedetti Diamond)
Professional football field (Candlestick Park)
Professional baseball field (Dodger Stadium)
Horse race track (37°53'06"N 122°18'40"W) – *Golden Gate Fields*
Airport (San Francisco International Airport)
Docks (Yusen Terminals, Oakland)
Railroad yard (37°48'05"N 122°17'59"W)
Ships (USS Hornet Museum)
Yacht club (37°48'27"N 122°26'02"W)
College (University of California, Berkeley, California) – *Notice the sports fields.*
Freeway overpass (37°44'09"N 122°24'26"W)

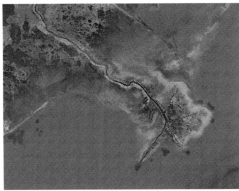

Mississippi Delta, Louisiana (Google Earth Image 2012 TerraMetrics, Data SIO, NOAA, U.S. Navy, NGA, GEBCO, Image NOAA, Image 2012 Digital Globe, 29° 9'32"N 89°13'53"W, eye alt 90 mi)

Finding Patterns in the Land

Out in the country, the airports, golf courses, and football stadiums are far apart, but the patterns in the land are fascinating. Agricultural practices around the globe sculpt the hills and organize the prairies into geometric shapes. Streams, which turn into rivers, which run into the sea, decorate the land. Islands come in all shapes, colors and sizes. And nature's signatures on the deserts make them works of art from above. Below are regions that have characteristic patterns. In each case, two or three locations are identified along with specific features to note.

Vegetation
Agriculture (Mt. Vernon, South Dakota, eye alt 10 mi) – *The fields are divided into one-mile sections and townships. More on this later! Also look at Atkinson, Nebraska, eye alt 10 mi, to see fields that are watered with center pivot irrigation systems, and Lancaster, Ontario, Canada to find fields aligned perpendicular to the river.*

Oak forests (37°31'N 122°25'W, eye alt 5000 ft) – *Fly around the hills south of San Francisco to see the difference between grassland, shrub land and forest.*

Orchard (36°16'30"N 119°40'37"W, eye alt 800 ft) – *Look for the rows of trees compared to the random distribution of a native forest on the hillsides south of San Francisco.*

Topical forests (8°09'S 60°45'W, eye alt 4000 ft) – *Zoom in until you can see the tops of the trees and the totality of the canopy cover. Then zoom out to appreciate the extent of the Amazon Forest.*

Clearcuts (38°24'N 120°19'W, eye alt 12 mi) – *Look for the lighter rectangular plots of land where clearcutting has occurred. Did your newspaper come from these forests?*

Water

Lake (San Andreas Lake, California, eye alt 10 mi) – *Use the Clock to go back to 5/2/11 and note that half the water in the lake is black and half is bright because part of the lake was imaged on a day when the Sun was not shining on the lake (dark), and part when it was (bright).*

Dammed lake (Lake Tawakoni, Texas, eye alt of 27 mi) – *Dammed lakes always have a straight line on one shore where the dam stops the flow of the stream or river. Zoom out to see other dams that are part of an elaborate water system.*

River (St. Louis, Missouri) – *Look for the wide Mississippi River. Use the Ruler to measure its width. Zoom in to see the bridges, and zoom out to see where the Missouri and Mississippi Rivers join. Zoom back in over St. Louis to an eye alt of 600 ft and travel south on the river to find some of the barges that are still popular for transporting goods.*

Lake lines (Mono Lake, California, 38°05'50"N 118°53'40"W) – *Mono Lake was once much deeper before its water was diverted to Los Angeles. Rings around the lake were formed as the lake level slowly dropped. Look along the shore of Lake Mead in Arizona and Nevada to see if this lake is at its fullest capacity, or if there are water lines from years when the lake was fuller. Note that the lake elevation in the Google Earth Status Bar stays constant as the true lake level rises and falls because the lake elevation was measured from space only once in the year 2000 on the Shuttle Radar Topography Mission on the Space Shuttle.*

Wetlands (The Everglades, Florida, 25°31'N 81°05'W, eye alt 10 mi) – *This is a small part of the Everglades – fly around southern Florida to see the diversity of vegetation in this wetland region. Note some of the elevations in southern Florida. As sea levels rise, what will happen to the Everglades?*

Oxbow lakes (Desoto Lake, Alligator, Mississippi) – *Oxbows are formed when a sharp meander in a river is cut off from the main channel, thus forming a lake. These meanders are most common where rivers cross flat topography. As the stream erodes the outer bank, it deposits sentiment on the inner bank thus creating a curve in the river.*

Deltas (29°08'N 89°15'W, eye alt 45 mi) – *My favorite is the Mississippi Delta – a bird-foot delta that spreads out into the Gulf of Mexico, like the webbed foot of a bird, dumping water and silt from across much of the United States. Search for St. Gabriel, Louisiana at an eye alt of 35,000 ft and follow the Mississippi south all the way down to the delta using the Move joystick. Notice the cropland, the city of New Orleans and the river traffic as you go. Other favorites are the Yukon Delta, Canada (63°N 164°W, eye alt 65 mi) and the Lena River Delta, Russia (72°59'N 126°26'E, eye alt 160 mi). One of the most dramatic is* the Betsiboca Delta in Madagascar (search for Mahajanga, Madagascar and look inland for the orange braided delta, eye alt 34 mi). This muddy delta is being formed by massive amounts of silt being washed into the ocean as the land all over Madagascar is cleared for grazing and farmland.

Waterfalls (Niagara Falls, New York) – Waterfalls can look impressive using the Look joystick to gain a perspective view. Use the Look and Move joysticks to dive over Niagara Falls, or Victoria Falls between Zambia and Zimbabwe in Africa. Use elevation in the Status Bar to find out which waterfall is the highest.

Deserts
Barchan dunes (16°42'29"S 71°50'23"W, eye alt 8000 ft) – East of La Joya, Peru; barchan dunes are formed where wind blows constantly from one direction.

Star dunes (31°13'15"N 7°53'25"E, eye alt 16000 ft) – Star dunes in the Grand Erg Oriental, Algeria are formed in regions with multi-directional winds. There are many other dune shapes formed by different wind patterns; wander around the Namib Desert, Namibia to see which way the wind blows.

Dry lakes (Bonneville Salt Flats, Utah, eye alt 135 mi) – Dry lakes were once filled with water. The current flat pans formed from salts and silt deposited as the lakes dried. These "lakes" are so flat that they are often used to set world speed records. Some dry lakes can be wet during the rainy months of the year. The largest dry lakebed in the world is Salar de Uyuni, Bolivia.

Alluvial fans (36°49'43"N 117°14'08"W, eye alt 12 mi) – Look for fan-shaped features spreading from the mountains into the valley in Death Valley. Alluvial fans are created as streams from the mountains carry sediments out onto the valley floor.

Geologic Structures
Fault lines – Draw a Path between 34°03'N 117°01'W and 34°48'N 118°40'W to see the lower section of the San Andreas Fault. Zoom in to see if the fault line is a sharp crack in the crust or a series of hills. Then search for Tomales Bay, California to see the upper section. The linear structure of the bay was created by the fault. Zoom out to see the full extent of the fault as it crosses California. Not all fault lines are so distinct.

Volcanos – Search for the Galapagos Islands, Ecuador and look for the craters at the centers of the island volcanoes, and the lava flows down the sides of the volcanoes. Use the Ruler to measure the distance between the islands that kept Darwin's finches isolated.

Canyons and sedimentary layers – To appreciate a canyon, search for "Grand Canyon National Park" and use the Look joystick and North-up button to look *around. Notice the Colorado River, the sedimentary layers along the walls of the canyon, and, using the elevation, the depth of the canyon.*

Folded mountains (40°35'N 77°23'W, eye alt 140 mi) – The Appalachian Mountains are one of the best examples of folded mountains. These mountains were formed as two tectonic plates collided and squished layers of sedimentary rock together. The more resistant layers remain as ridges and the softer layers eroded away. Zoom out to see that the Appalachians were a significant barrier to westward movement when America was first settled. The mountains were formed when Pangaea became a continent 480 Mya. Because North America and Africa were then both part of Pangaea, the Little Atlas Mountains in Morocco (29°N 8°W, eye alt 100 mi) were once part of the same mountain range as the Appalachians. Another excellent example of folded mountains is the Macdonnell Range in Australia – search for 24°15'S 132°07'E, eye alt 50 mi.

Mines – Mines are often recognized first by their colorful tailings ponds. Look for a diamond mine in Orapa, Botswana, (21°18'26"S 25°22'11"E, 20,000 ft), a copper mine in Bisbee, Arizona (31°26'01"N 109°54'05"W, eye alt 14000 ft), and the Super Pit Gold Mine in Australia (eye alt 20,000 ft)

Finally, astronauts are always looking for the Richat Structure in Mauritania (you can find it using its name but be sure to zoom to an eye alt of 20 mi to appreciate its structure). What do you think it is? The astronauts call it Earth's bull's eye.

<u>Islands</u>
Continental islands (Trinidad and Tobago, eye alt 1200 mi) – *Continental islands lie on continental shelves. Trinidad and Tobago lie on the continental shelf of South America. You can see the extent of the continental shelf by noting the lighter shades of water in the ocean north of Venezuela.*

Barrier island (Cape Hatteras, North Carolina, eye alt 108 mi) – *A barrier island forms from sand deposited on the floor of a continental shelf. It was on Cape Hatteras, on Kill Devil Hills, that the Wright Brothers made their first flight.*

Volcanic arc islands (The Aleutians Islands, Alaska, eye alt 1000 mi) – *Volcanic arc islands form as one tectonic plate is subducted under another. In this case, the Pacific Plate is subducted under the North American Plate forming the long chain of volcanoes that make up the Aleutians.*

Volcanic islands from hot spots (Hawaii, eye alt 500 mi) – *Some volcanic islands form over a hotspot in the Earth's crust, or a region in the ocean where magma reaches the surface. Hawaii is a perfect example of this. Kauai was the first island to form over the Hawaiian Hot Spot. As Earth's crust moved to the northwest, Kauai moved away from the hot spot and O'ahu began to form over the hot spot. Now Kauai is the most eroded and vegetated because it is the oldest, and the Big Island of Hawaii shows the most evidence of fresh volcanic activity because it was formed most recently.*

Atolls (Wake Island) – *Atolls are coral reef islands formed when a volcanic island subsides leaving behind a circular island with a shallow lagoon in the center. Because the center is shallow, it is light turquoise in color.*

Ice
Snow (Mont Royal Park, Montreal, Canada, eye alt 7300 ft) – *Ice on Earth is almost always white, but comes in many shapes, sizes and textures. The most familiar ice to most of us is snow. Google Earth seeks to provide images that are free of snow, so snow can be difficult to find, but Mont Royal Park is one example. You will need to use the Clock to find the snow day.*

Alpine Glaciers (Mer de Glace, France, eye alt 50,000 ft) – *Snow that does not melt year after year because summer temperatures do not rise above zero creates glaciers. Glaciers can be found in both high latitudes and high altitudes. The interesting thing about glaciers is that they move – very slowly. The snow piling up on the top of the glacier gets so heavy it pushes the ice formed in previous years down the mountain like a very slowly flowing river. Glaciers melt as they flow down the mountain because temperatures get warmer at lower elevations. Glaciers that melt before they reach the bottom of a mountain are called alpine glaciers. Mer de Glace Glacier terminates about a mile and a half east of the town of Mer de Glace. The stripes on Mer de Glac are called ogives and are created at an icefall, where the glacier seasonally flows over a steep change in topography (like an icefall, probably at about 45°53'05"N 6°56'04"E). Once formed, the ogives flow down the mountain in bands. Since the ice flows faster in the middle of the glacier, the ogives become more curved as they flow down the mountain. Another famous alpine glacier is Fox Glacier, New Zealand at 43°30'40"S 170°05'59"E, eye alt 21000 ft.*

Tidewater Glaciers (Eqalorutsit, Greenland; 61°20'N 45°47'W) – *Glaciers that reach the sea before melting are called tidewater glaciers. This tidewater glacier in Greenland reaches for the Labrador Sea through a long fiord. Also look at Upsala, a beautiful tidewater glacier on the Argentina/Chile border at 49°55'S 73°17'W, and Bear Glacier, Alaska at 59°58'N 149°36'W, a famous site in Glacier Bay.*

Moraines (Malaspina Glacier, Alaska; 59°56'N 140°30'W, eye alt 50 mi) – *Along with looking for the ice of a glacier, look for moraines, or piles of debris alongside (lateral) or ahead of (terminal) the glaciers. These piles of debris remain as the glacier retreats. Long Island, Martha's Vineyard and Nantucket Island are giant moraines left behind by the glaciers of the last ice age!.*

Icebergs (61°15'N 45°52'W, eye alt 8000 ft) – *Once glaciers reach the sea, chunks of ice fall off into the ocean. These are called icebergs. They float around for a while until they melt. Use the Ruler to determine the size of some of these icebergs*

Sea ice – *Sea ice is frozen ocean water that floats on the surface of the ocean. In the center of the Arctic Ocean, sea ice remains all year and into multiple years. At the edge of Arctic Ocean at lower latitudes, the sea ice melts each summer and refreezes in the winter. Chunks of sea ice that break apart from the main pack are called ice floes (75°56'S 149°55'20W, eye alt 25 mi). When sea ice is connected to the shore, it is called fast ice (74°10'S 126°55'W, 50 mi).*

Borders

Borders between countries and states are sometimes defined by natural features, like mountains and rivers, and sometimes by lines of latitude or longitude. The Texas/Mexico border, for example, is defined by the Rio Grande, and much of the Canada/United States border is along the 49th parallel. Some borders can be seen from space because land-use practices in two countries separated by a border are very different. Search for 32°39'22"N 115°36'46"W at an eye alt of 44,000 ft. The border between California and Mexico runs east/west across the 3D Viewer. Notice the difference in agricultural practices. The fences people build to protect borders can also often be seen on Google Earth:

Dingo Fence (29°00'S 141°01'E, eye alt 1500 ft) – *This 3,488-mile long fence was built to keep dingos (a kind of wild dog) out of the fertile southeast portion of Australia where sheep are raised.*

Great Wall of China (39°48'05"N 98°12'58"E, eye alt 10,000 ft) – *The westernmost end of the Great Wall is near Jiayuguan, China. The square-shaped structure is a gate in the wall called the Jiayu Pass. The Wall is to the northeast and southwest of the gate. Another place to see the wall is at Badaling (40°21'08"N 116°00'21"E, eye alt 4,000 ft).*

Xian, China – *Zoom to an eye alt of 20,000 ft and look for the wall around the city. Built during the Ming Dynasty as a defensive wall, it survives today. Zoom in further to examine this impressive wall that has lasted so many years.*

Shadows

Since structures are only photographed from above on Google Earth, shadows can sometimes help identify a feature. Here are a few examples.

Dino Park, Furth, Germany (49°28'50"N 11°00'07"E, eye alt 1200 ft) – *How big are the dinosaurs compared to the size of your house?*

Camels in Chad, Africa (15°17'40"N 20°28'48"E) – *Keep zooming in to see these graceful creatures.*

Eiffel Tower, Paris, France – *Use the Clock to discover that the shadow is never on the south side of the tower because the Sun always shines from the south for structures in the northern hemisphere.*

The Obelisk of Buenos Aires, Buenos Aires, Argentina – *In this case, the shadow is never on the north side because the Sun always shines from the north in the southern hemisphere.*

London Eye, London – *Click through the Clock to see the shadow at different times of day.*

Segovia Aqueduct, Segovia, Spain (eye alt 4000 ft) – *Set the Clock slider to 8/2/2007 to appreciate what the Romans built by observing its shadow.*

Monterey Bay, California (Google Earth Image, Data LDEO-Columbia, NSF, NOAA, Image 2012 TerraMetrics, Data SIO, NOAA, U.S. Navy, NGA, GEBCO, 36°44'06"N 122° 5'14"W, eye alt 90 mi)

Diving Into the Ocean

The treatment of the ocean in Google Earth is different from that of land. A high-resolution global image of the ocean would require a lot of image storage space, and the ocean is different every day! So most of the ocean surface in Google Earth is an animation, or false surface, and is colored according to the ocean's depth: the darker the blue, the deeper the ocean. Elevation (elev) in the Status Bar over the ocean is the depth of the ocean relative to sea level, rather than height above sea level, as it is over land. Eye alt over the ocean is the height of your eye above sea level as it is for land. Search for Guam and zoom out so most of the Pacific Ocean is in view. Notice the very dark arc above Guam – this is the Mariana Trench; the deepest part of the ocean. As you slide your cursor over the trench, note the elevation!

Search for Monterey Bay, California (eye alt 110 mi) to explore an underwater canyon. Make sure the "Water Surface" under View in the Google Earth Menu

Bar is checked on, and slowly zoom in over the bay to about 10 mi eye alt. You should start to see artificial waves marking the surface of the ocean. Click the Water Surface on and off to see the difference. Leave the Water Surface on and use the Look joystick to tilt your perspective – the water surface will become more apparent. Return to a view from above the bay (click "u" on your keyboard) and slowly drop down under the water's surface. Then use the Look joystick, including the North-up button, to look around the canyon where sea otters play.

Search for Nikumaroro Island, Kiribati where Amelia Earhart is thought to have crashed to see what real imagery of the ocean's surface looks like. Surrounding Nikumaroro Island is an area of the ocean that is part of the image of the island. Look for the transition between the *image* of the ocean and the *animation* of the ocean. Look for waves and currents on the ocean's surface near the island. More of the ocean will be explored when you dive into the Layers Panel in the next section.

Himalayas (Google Earth Image, 2012 Cnes/Spot Image, Image 2012 TerraMetrics, Data SIO, NOAA, U.S. Navy, NGA, GEBCO, 30°23'44"N 85°12'11"E, eye alt 1300 mi)

Remembering the Global View
It is easy to focus on higher and higher resolution imagery, but don't forget about the global view. Here are a few regions that might inspire you.

Amazon Forest (4°S 64°W, eye alt 2,500 mi) – *Notice the Amazon forest – the darkest green region of northern Brazil. The combination of warm equatorial temperatures and heavy rainfall make Brazil an ideal place for tropical forests. It would take a lot of cutting to clear this entire forest. But zoom in around Rondonia, Brazil and see the progress!*

Land bridge between Asia and North America (64°N 170°W, eye alt 3000 mi) – *Notice how shallow the ocean is between Alaska and Russia. When much of the world's water was locked up in glaciers during the last ice age, this continental shelf was likely above water allowing people to migrate from Asia to North America.*

Migration – *Gray whale migration happens every year from Baja California, Mexico to the continental shelf surrounding Alaska. The shelf is conducive to ocean upwelling, which brings nutrients to the surface that krill like to feed on. And gray whales love krill. Zoom out to 45°N 131°W, eye alt 4000 mi and follow the whale's migration path from Baja to Alaska. Zoom in to the coves (eye alt 30 miles and closer) in Baja California where the whales give birth in the winter months: Laguna Ojo de Liebre (27°51'44"N 114°14'04"W), Laguna San Ignacio (26°49'24"N 113°12'43"W) or Laguna Magdalena (24°35'57"N 111°56'02"W).*

Sahara Desert – *You might think of the Sahara Desert as a vast sea of tan sand, but use your navigation tools to look at the textures and colors created by the desert landforms, including sand seas (called ergs, like Grand Erg Oriental, Algeria), stone plateaus (called hamadas, like Tassili N'Ajjer, Algeria at eye alt 500 mi), wadis or dry valleys (Wadi al-Hayat, Libya, for example – zoom in to see the streambeds), chotts or salt flats (Chott Melrhir, Algeria, for example), and the occasional oasis (like Ouargla, Algeria; zoom in to see if anyone lives on this oasis, and zoom out to see how isolated the oasis is in the middle of the Grand Erg Oriental).*

Panama Canal, Panama – *The canal cuts through Central America in the country of Panama. Zoom out until North and South America are in view and think about the challenges of traveling by ship from New York City to San Francisco without this canal. Zoom in to see how much of the crossing is man-made locks and how much is natural lakes. Use the elev to help you find the locks.*

Suez Canal, Egypt – *This "Highway to India" connects the Mediterranean and Europe to the Red Sea and Asia. How far did ships have to travel to carry goods between these two continents before this canal was carved? Are there any locks on this canal?*

Himalayas (30°23'N 81°55'E, eye alt 1500 mi) – *This massive range was formed (and continues to be formed) as the Indian Plate slowly crashed into the Eurasian Plate. Sedimentary layers from the seabed that once separated the Eurasian and Indian plates have been pushed up forming the highest peaks in the Himalayas. Drag your cursor from India across the Himalayas to the Tibetan Plateau and watch the elevation change. Look for the Indus and Brahmaputra Rivers, which drain from the glaciers in the Himalayas into the Arabian Sea and the Bay of Bengal. Water from these and other Himalayan rivers support a large fraction of the world's population.*

Alaska patchwork of images (Google Earth Image 2012 TerraMetrics, Data SIO, NOAA, U.S. Navy, NGA, GEBCO, 58°24'03"N 155°47'08"W, eye alt 75 mi)

Enjoying a Smoothed Earth
Not long ago when you looked at a global or regional view in Google Earth, it looked like a patchwork quilt. This is because Google Earth is made up of thousands of images that are mosaicked together like a puzzle. These images are taken by a variety of cameras and instruments, and are taken in different seasons and under different sun angles. So the colors of each image are slightly different and, when stitched together, they look patchy. The current Google Earth globe is covered in images that have been smoothed together to make a seamless looking view. But, if you click on the Clock over a region, you will again see image boundaries.

A side effect of mosaicking images from different view angles is that buildings lean in different directions. Search for Singapore at 1°16'53"N 103°51'07"E (eye alt 1200 ft) and click through the Clock slider to see the buildings tip north and then south. For example, on 4/6/2010, the satellite was to the southwest of the buildings, so the buildings look like they are leaning to the northeast.

Snow over the Kremlin, Russia (Google Earth Image 2012 DigitalGlobe, 55°45'07"N 37°37'26"E, eye alt 3500 ft, Clock 3/13/2010)

Discovering a Dynamic Earth
You may have already noticed that you rarely find images with snow, clouds, smoke from fires, or ash plumes from volcanic eruptions. The top-layer Google Earth global image tends to be cloud-, smoke-, and snow-free so you can see

the surface. When you turn on the Clock, you can sometimes find images with these more dynamic features. Here are a few I have discovered. Note that it is often easy to confuse clouds and snow as they both appear as white blobs from above.

Clouds
Galapagos Islands, Ecuador (0°52'S 91°02'W, eye alt about 25,000 ft) – *Look for the cloud shadows as a clue that these are clouds and not snow.*

Snow
Mont Royal Park, Montreal, Canada – *Use the Clock to look for snow, summer green leaves, and fall red and orange leaves.*

Roseville, Minnesota – *Look for the snow in 2005. The image looks like it is black and white, but zoom in and you will see color around the snow.*

The Kremlin, Moscow, Russia – *Watch the snow and leaves on the trees come and go in the parks surrounding the Kremlin.*

Mt. Etna, Sicily – *Use the Clock to look for snow and small ash plumes (3/22/10).*

Fires
Alnwick, England (at 55°24'31"N 01°39'55"W) – *Look for the crop fire – you may need to use the Clock.*

Although it is rare to see active fires in Google Earth imagery, the USDA Forest Service tracks fires in real-time using Google Earth. Go to http://activefiremaps.fs.fed.us/googleearth.php and click on "Fire Detections: Current". You will be returned to Google Earth over the United States. Zoom in until you see a number of yellow and red dots. Zoom in on one of the dots (below 20 mi eye alt) and you will see a series of squares and a legend in the upper left of the 3D Viewer. The squares are colored according to the most recent time an instrument named MODIS on a NASA satellite detected a fire in that pixel. These maps are extremely valuable if there is a fire in your region.

My kmz File

You might already be tired of typing lat/longs, or, if not, I guarantee you will get tired in the next section. You have the option of downloading *my* kmz file, which follows the sections in this book. To do this, go to http://googlingearth.com and click on the round Earth at the bottom of the web page. My folder will appear in your [Temporary Places] folder; it is called [googlingearth.kmz]. If you want to keep it, move it to *your* [My Places] folder.

Digging into Layers

The image view of Earth provided by Google Earth tells you a lot about a location or region, especially if you have already visited or studied that place. If you want to learn more, you can add layers of information on top of the image using tools in the Layers Panel and under View in the Google Earth Menu Bar. Below are some of my favorite layers that hopefully will serve to encourage you to explore more layers. Note that the layers often change; sometimes if a layer disappears, you can find it in the Earth Gallery (more on that below).

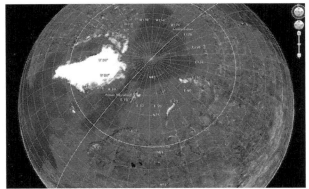

North Pole with Grid (Google Earth, 2012 Cnes/Spot Image, Image 2012 TerraMetrics, Data SIO, NOAA, U.S. Navy, NGA, GEBCO, Image IBCAO, 77°15'05"N 47°15'26"E, eye alt 4500 mi)

Latitude Longitude Grid

Starting with a global view in the 3D Viewer, look in the Google Earth Menu Bar for View and choose "Grid". Latitude and longitude lines will appear on your globe. Use the cursor or the Move joystick to roll around Earth noting how the lines of latitude and longitude are different at the equator and at the poles. Search for Greenwich, England to verify that it marks zero degrees longitude, or the Prime Meridian.

Google Earth Overview Map

Scale Legend and Overview Map

You can add a Scale Legend and an Overview Map of the world to the 3D Viewer to help reference your location by clicking on "Scale Legend" and "Overview Map" under View in the Google Earth Menu Bar. The Overview Map has red crosshairs on it to show your current location. The Scale Legend length changes with eye alt.

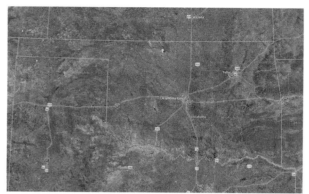

Oklahoma bordered by latitude 37°N along the north border and the Red River along the south border. (Google Earth Image 2012 TerraMetrics, 2012 Google, 2012 INEGI, 35° 4'54"N 98° 2'10"W, eye alt 540 mi)

[Borders and Labels], [Places], [Photos] and [Roads]

Click off the Grid and navigate back to the Golden Gate Bridge by double clicking on your Placemark. Zoom to an eye alt of about 30,000 ft. Look in the lower left of the Google Earth Window for the Layers Panel. There is a little arrow just to the left of the words identifying the panel; be sure it is turned down so the layers in the panel are visible. If some of the layers are out of view, you can close the Places Panel by turning that arrow sideways to make more room for the Layers Panel.

Look for the first layer called [Borders and Labels]. Click it on by checking the box to the left of the layer and note the labels that have appeared in the 3D Viewer. Zoom out to look for the county borders, the border of California, and then the countries in North America. Expand the [Borders and Labels] layer by turning the arrow down and clicking off all layers except [Borders]. Expand the [Borders] layer and click off all layers here except [Coastlines] to appreciate how you can manipulate these features.

Click on the [Roads] layer and zoom in slowly over San Francisco to see the interstates, then the highways, then the streets and roads. Finally, click on the [Places] layer to see that many places are marked with little icons. You can find out what places are and something about them by clicking on these icons.

Return to the Golden Gate Bridge, zoom to at eye alt of 30,000 ft, and click off the [Borders and Labels], [Places] and [Roads] layers. It is helpful to click off each layer when you are finished with it, or Google Earth will slow down significantly. Expand the [Photos] layer to see the [Panoramio] and [360 Cities] layers. Click on the [Panoramio] layer to see icons marking photos taken by people all around the world. Click on a few to get an idea of what these photos offer. Sometimes the name of the photo can help you identify a location that is not marked in [Borders and Labels] or [Places]. You can upload your own photos using http://www.panoramio.com.

Click off the [Panoramio] layer and click on the [360 Cities] layer. Click once on one of the 360 City icons to see a popup window with a photo; click on the photo to fly into a panorama. Once in the panorama – which is a photograph taken from the ground, use the joystick in the upper right of the 3D Viewer to move around. When you are done looking around, click on "Exit Photo".

3D view of the Palace of Fine Arts, San Francisco, California (Google Earth Image Data LDEO-Columbia, NSF, NOAA, Image 2012 TerraMetrics, Data SIO, NOAA, U.S. Navy, NGA, GEBCO, Image 2012 DigitalGlobe, 37°48'10"N 122°26'54"W, eye alt 300 ft)

[3D Buildings]
Search for the Palace of Fine Arts, San Francisco, California, zoom in to an eye alt of about 3000 ft, and click on the [3D Buildings] layer. Use the Zoom and Look joystick to investigate the palace. Can you fly into it? Note that it is indeed three-dimensional. These layers are models of structures, not images. Most of the famous places around the world have [3D Buildings] models, and more are being created all the time. Notice also that there are 3D trees around the Palace of Fine Arts. Expand the [3D Buildings] layer to see that you can click the trees on and off.

Navigate to the Golden Gate Bridge and investigate its [3D Buildings] structure. Look in the water below the 3D bridge to see the image of the bridge taken by the satellite. Since all the satellite imagery is taken from above, and the imagery is draped over a 3D topographic model, the bridge in the image becomes part of the water. Below are some of my favorite 3D structures.

Neuschwanstein Castle, Germany
Angkor Wat, Cambodia
Forbidden City, Beijing, China
Windsor Castle, Windsor, England
Chateau de Chambord, Chambord, France

Statue of Liberty, New York
Notre Dame, Paris, France

Sea lion tracking off California (Google Earth Image 2012 TerraMetrics, Image U.S. Geological Survey, 33°20'49"N 119°31'43"W, eye alt 32 mi)

[Ocean]
The layers in the Ocean are constantly changing as Google Earth develops this feature. To get an idea of the layers that are available now, click on the [Animal Tracking] layer under the [Ocean] layer and look for the whale, the seal or one of the fish. Click on each animal to learn more about it and then click on "Download Track" in the popup window to see where that animal has traveled.

Hurricane Sandy (Google Earth Image, Data LDEO-Columbia, NSF, NOAA, Image 2012 TerraMetrics, Data SIO, NOAA, U.S. Navy, NGA, GEBCO)

[Weather]
The [Weather] layer provides up-to-date information on clouds, rain, and temperature around the world. The most entertaining use of this layer is when a hurricane is traveling across the ocean. If there is a hurricane, click on the [Clouds] layer and look for the hurricane. Placemark it's eye, or center, and name the Placemark with the date. Add it to a new folder named after the hurricane. Each day mark a new Placemark and you will be able to track the hurricane's path. Watch for the formation of the eye!

Nile night lights. (Google Earth Image U.S. Geological Society, 2012 Cnes/Spot Image, Data SIO, NOAA, U.S. Navy, NGA, GEBCO, 27°47'42"N 31°25'44"E, eye alt 1200 mi)

[Earth City Lights]

The Google Earth globe is always lit unless the Sunlight tool is turned on. Zoom to a global view and look in the Layers Panel under [Gallery] → [NASA] → [Earth City Lights] and click open this layer. The Earth is now totally dark. The images that make up this layer were collected at night and mosaicked together to form a global view of the Earth at night highlighting city lights. Navigate around to see where population centers are located. Compare the eastern and western United States, eastern and western China, and North and South Korea Note that you can add the [Borders and Labels] layer to determine the locations of these countries. Take a close look at the Nile River region across Egypt; the lights highlight how important the Nile is to the people of Egypt. Navigate around the Persian Gulf – perhaps oil platforms like the one at 26°35'36"N 52°00'05"E are responsible for some of these bright lights. Click off [Earth City Lights] and zoom in to see the oil platform. Look in the channel between Japan and South Korea near the island of Cheju for the bright greenish lights. These are squid fishing boats that work in the night using bright searchlights to attract the squid.

Elephants in Chad. (Google Earth Image 2012 GeoEye, 10°54'13"N 19°55'59"E, eye alt 1500 ft)

[Africa Megaflyover]

In 2004, a conservationist named Mike Fey traveled around Africa in a little red Cessna airplane and photographed villages and animals. His photographs have been added to Google Earth as a layer. Click open the [Africa Megaflyover] layer under [Gallery] → [National Geographic Magazine], and navigate to Africa. Zoom in until little red airplane icons appear. Choose one and continue to zoom in on it until you see an area with a very high resolution photograph; usually about 2500 to 1000 ft eye alt. As you look at more of these high-resolution photos under the little red airplane, you will find large African animals and small African villages. Some of my favorites are:

3°13'33.61"S 35°02'29.71"E – *You can see the shadow of the little red airplane in Kenya.*
Elephants in Chad (10°54'12.67"N 19°55'55.85"E)
Camels in Chad (15°17'40.24"N 20°28'47.86"E) – *Note their shadows.*
Bazaar in Tanzania (3°00'7.97"S 33°05'24.65"E)
Buffalo and gazelles in Chad (16°24'15.93"N 19°54'49.86"E)
Hippos in Zambia (13°42'7.09"S 31°8'29.21"E)
Hippos in Tanzania (6°37'46.10"S 31°08'12.95"E)
Buffalo and egrets in Mozambique (18°43'40.52"S 35°55'57.98"E)
Flamingos in Mozambique (21°50'36.23"S 35°27'00.48"E)
Lechwe in Zambia (15°50'20.40"S 27°11'50.41"E)
Seals in Namibia (18°26'45.50"S 12°00'44.20"E)
Oryx in Namibia (24°57'18.52"S 15°51'30.56"E)
Cattle in Botswana (19°34'28.21"S 22°13'48.00"E)
Village in Niger (13°47'21.49"N 9°00'02.09"E)
Village in Chad (13°35'42.13"N 20°00'23.61"E) – *Look for the people looking up!*

Rumsey map of San Francisco, California. (Google Earth Image, Data SIO, NOAA, U.S. Navy, NGA, GEBCO, Image 2012 TerraMetrics, Data CSUMB SFML, CA OPC, 37°47'08"N 122°24'25"W, eye alt 36 mi)

[Rumsey Historical Maps]

In addition to historical images, Google Earth offers overlays of historical maps in some locations. Zoom to an eye alt of about 30,000 ft over San Francisco and look in the Layers Panel under [Gallery] for [Rumsey Historical Maps]; click it on. David Rumsey is a map collector. Many of his maps have been digitized and added to Google Earth. Additional maps are available on his website: http://www.davidrumsey.com/. Look for the yellow compass roses in the 3D Viewer and use your cursor to identify the "San Francisco 1859" rose. Click on it and a popup window will appear. Click on the map in the popup window and the window will disappear and a map will appear in the 3D Viewer. Zoom out to see the area covered by the map, and zoom in to see the detail in the map. Look in the Places Panel in the [Temporary Places] folder for a folder called [North_America/US_Cities/San_Francisco_1859]. If you click the folder on and off, you can see the Google Earth image alternately with the map. Look for Mission Dolores, Yerba Buena Island, the Golden Gate Bridge and Alcatraz. How has San Francisco changed over the years?

Other Interesting Historical Maps:
Washington DC: Washington DC 1851
New York City, New York: New York City 1836, 1851 or 1852
Rome, Italy: Ancient Rome 1830
Israel: Egypt Palestine 1818
Mexico City, Mexico: Mexico City 1883

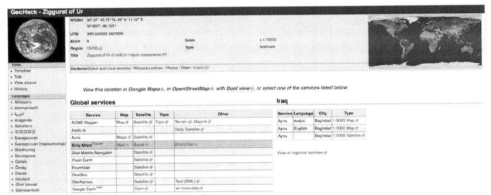
Wikipedia link to Google Earth.

[Wikipedia] Backward and Forward

If you happen to search for a place and wonder what it is, click open the [Wikipedia] layer in the [More] layer and look for the Wikipedia icons. If there is one where you are (you may need to zoom in or out a bit; and there may not be one), click on it to learn more. For example, search for 41°54'13"N 12°28'00"E at an eye alt of 2000 ft. What is the building with the geometric garden wall? Click on the [Wikipedia] layer and look for the Wikipedia icon to discover it is Castel Sant'Angelo in Rome, Italy. Have you read <u>Angels & Demons</u> by Dan Brown? You will usually see some photos in the Wikipedia popup window. To learn even more, look in the popup window for "Full Article", click on it, and you will enter the Internet Viewer and be able to read the full Wikipedia article. Click on <<Back to Google Earth to return to the 3D Viewer.

If, on the other hand, you know about an interesting place but are having trouble finding it, search for it using Wikipedia on your Internet browser. As an example, search for "Ur Ziggurat", an ancient Sumerian temple in Iraq, on the Internet on Wikipedia. Look in the right hand column of the first Wikipedia page for a latitude and longitude. You can copy this lat/long and paste it in the Google Earth Search Panel, or, even cooler, you can click on the lat/log and you will be taken to a new web page that has links to all kinds of graphic viewers. Look in the table in the "Service" column for "Google Earth", and click on "Open". You will automatically be taken to Google Earth and then right to the Ur Ziggurat (after a few questions in popup screens that you should be used to by now).

Kennesaw Mountain National Battlefield. (Google Earth Image 2012 Google, 33°57'38"N 84°35'45"W, eye alt 1300 ft)

[US National Parks]

Are you planning to visit a National Park? You can find out all about the roads, trails and sites using Google Earth. Look in the Layers Panel for [More] → [Parks/Recreation Areas] → [US National Parks]. Click this layer on for all options and search for your favorite park; Yosemite National Park, for example. Trails are highlighted in red, park boundaries in green (zoom out to see them), and visitor facilities use different icons like tents for camping areas.

It is likely that even more information will soon be available for US National Parks on Google Earth. Using your Internet Browser, search for Kennesaw Mountain National Battlefield. Look in the left of the top web page for "Maps>" and click on it. Look for 'Kennesaw Mountain NBP Google Map" and click on it. Look on this page for "Download the Kennesaw Mountain Google Earth Map" and click on it. A very detailed map and features will appear in your [Temporary Places] folder. To appreciate this information, click off everything in the [Kennesaw Mountain National Battlefield Park] layer in the [Temporary Places] folder, and open (but do not check) the [Confederate forces] folder. Then click on the [Confederate's Gun Positions] layer. Do the same for the [Federal forces] → [Federal's Gun Positions] to see how the two armies lined up for battle.

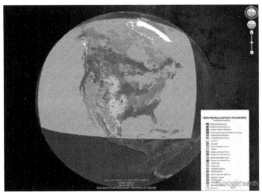

North American Land Cover Characteristics. (Google Earth, Data SIO, NOAA, U.S. Navy, NGA, GEBCO, Image Landsat, Image IBCAO, Dartmouth Flood Observatory, University of Colorado, 37°39'58"N 95°28'10"W, eye alt 6835 mi)

Earth Gallery

At the top right of the Layers Panel is a button called Earth Gallery. Click on it and you will see the Gallery in the 3D Viewer, which has now turned into an Internet Viewer. The layers are divided into several categories as listed under the "Explore" button. Click on a map and it will display within the Internet Viewer on Google Maps. To get it on Google Earth, click the button at the lower right of the map. It will appear in the Layers Panel but will disappear when you close Google Earth. Sometimes the map does not appear in Google Earth and sometimes you need to zoom in or out to find it. You can also search for layers in the Gallery by typing your search into the Google search bar at the top of the Earth Gallery page.

Google Earth Flight Simulator airplanes.

Flight Simulator

If you want to get seasick, you will be glad to know that Google Earth includes a flight simulator. Go to Tools in the Google Earth Menu Bar and choose "Enter Flight Simulator". In the popup window, choose the SR22 (the slower of the two options), "Current View", and click on "Start Flight". You will see a screen with some heads-up-displays (HUDs) and you will already be flying. Press the space bar to stop, and again to continue flying. You will be able to control your plane (or crash your plane) using buttons on your keyboard. The most important ones are:

Increase Thrust	Page Up
Decrease Thrust	Page Down
Left Brake	, (Comma)
Right Brake	. (Period)
Elevator Push	Up button
Elevator Pull	Down button
Rudder Left	Shift+Left Arrow
Rudder Right	Shift+Right Arrow
Center Aileron and Rudder	C
Extend or Retract Landing Gear	G
Toggle HUD on/off	H
Increase Flaps	F
Decrease Flaps	Shift+F
Rotate Viewpoint	Ctrl+Arrows for fast or Alt+Arrows for slow

For more controls, go to http://support.google.com/earth → Advanced features → Advanced user guide → Flight simulator.

Adventures in Google Earth

Now that you know how to use the Google Earth tools and know what the Earth looks like from many scales, it is time for some adventures. This section uses the Google Earth imagery and tools to explore a variety of subjects. The goal is to give you an idea of what *you* can do with Google Earth.

Exploring Washington DC

The White House (upper left) and the Capitol (right) connected by Pennsylvania Avenue in Washington DC. (Google Earth Image 2012 Commonwealth of Virginia, 38°53'30.37"N 77°01'31.19"W, eye alt 9500 ft)

In the late 1700's, America had won her independence, a constitution had been written, and a president had been elected. The capital of this new country was in Philadelphia, but the Founding Fathers wanted to establish a capital that was independent of any state. The Residence Act established the capital be on the Potomac River between the Eastern-Branch Potomac (today this is Anacostia River which meets the Potomac just south of Washington DC), and the Conococheague (today Conococheague Creek meets the Potomac north of Williamsport, Maryland – look or the zigzag of the stream). The exact location was to be selected by George Washington. Placemark the two rivers, along with Mt. Vernon, Virginia (George Washington's home; search for "George Washington's Mt. Vernon Estate") to see if Mt. Vernon's location may have influenced Washington's decision by its proximity. Washington DC is also downstream of the Potomac's fall line allowing navigation from the Atlantic to the capital.

Pierre Charles L'Enfant was selected to design the new capital. The layout of the city is based on north-south streets that are numbered and east-west streets that are lettered. Running diagonally are streets named after the original 13 states, with Pennsylvania Ave. running between the White House and the Capitol. Circular and rectangular parks sit at many of the intersections. Note that the White House and the Capitol, two of the three branches of government, are part of the overall design of the city. This is clear when the city is viewed at an eye alt of about 18,000 ft. But at the time Washington DC was built, the Supreme Court was in the Capitol building and was viewed as a federal court that settled disputes between the states. As the Supreme Court gained more power to rule on the constitutionality of laws and enforce the Bill of Rights, it moved to its own building just behind the Capitol. Find the White House, the Capitol and some of these famous building, all of which can be found by simply typing in their name in the Search box. Click on the [Rumsey Maps] layer, choose the 1851 map, and check to see that all 13 original colonies are represented in the first streets of Washington. Which buildings were built by 1851? With the Rumsey map layer on, you can see that some of present-day Washington was part of the Potomac River. Here are some other sites to investigate:

Washington Monument
Lincoln Memorial
Reflecting Pond
Jefferson Memorial
Vietnam Veterans' Memorial
Korean War Veterans' Memorial
World War II Memorial (38°53'22.31"N 77°02'26.59"W)
Smithsonian National Museum of Natural History
National Air and Space Museum
Library of Congress
Ford's Theater
Martin Luther King Jr. National Memorial – *Use the Clock to watch the construction.*
The Pentagon, Virginia
Naval Observatory, Washington DC – *Also home of the Vice President.*

Measuring America

Surveyed land around Mt. Vernon, South Dakota showing townships, sections, and quarter and quarter-quarter sections. (Google Earth Image USDA Farm Service Agency, Image 2012 GeoEye, 43°42'37.93"N 98°15'37.20"W, eye alt 2800 ft)

The Revolutionary War left the country broke and setters headed west looking for land. The Founding Fathers realized that the solution to solving the country's financial woes was to sell land that was to be settled as laid out in The Ordinance of 1785. To sell the land, however, it had to be measured and documented. How to do that led to much debate. Should the land be divided into sections using a decimal system, or into sections that are more easily divisible by four. In the end, a system based on sections divisible by four was agreed upon, and the measuring began. Starting in East Liverpool, Ohio, one by one mile sections of land were surveyed. Every six by six sections would be a township, and a town would be established in every other township. Much of the surveyed land was flat and surveying was relatively easy. Other areas where hilly and forested which created obstacles making surveying more challenging. The result of this unprecedented survey created a patchwork quilt across America west of the Ohio River. Search for Mt. Vernon, South Dakota at an eye alt of 3000 ft to begin exploring the patterns on the land created by the surveyors. Not all farmers were prepared to work a one-mile by one-mile section of land, so quarter sections and quarter-quarter sections were also established.

The southern side of each section was aligned with lines of latitude. The north/south sides were aligned with lines of longitude. This system worked well except for the fact that lines of longitude converge as they head north. Corrections were made for this convergence by adjusting the north/south lines every 24 miles, or every four townships. To see what a correction looks like, follow the road that borders the west side of Mt. Vernon north for about 10 miles. The country road jogs at the intersection. Then fly 24 miles farther north to see the next correction. If you grew up in this part of the country, a correction would be a common reference point.

Looking at the land west of the Ohio River today using Google Earth suggests that the surveyors of the early 1800s were an impressive bunch. Comparing land established before the Revolutionary War, like the region surrounding the Cowpens National Battlefield in South Carolina, to that settled after surveying illustrates two dramatically different ways to divide up the countryside.

Planting Circular Fields

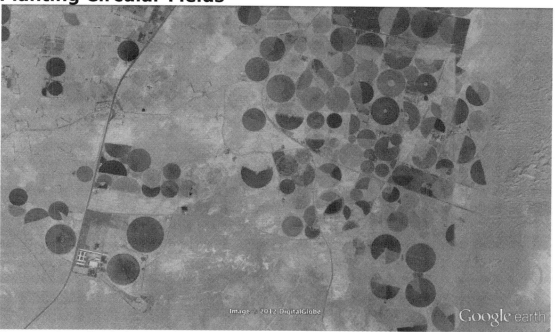

Center pivot fields in Saudi Arabia. The larger fields are a half-mile in diameter. (Google Earth Image 2012 DigitalGlobe, 23°51'41.36"N 47°06'40.53"E, eye alt 40000 ft)

I was showing slides from one of my Space Shuttle missions to a group of students. On the screen was a picture of circular agricultural fields in Saudi Arabia and I asked the students to tell me why these fields are circular. One student raised his hand and suggested they are circular to make the Earth look prettier for the astronauts.

In reality, circular agricultural fields are watered with a center pivot sprinkler system that feeds from a well at the center of the field. Most of the fields are one mile in diameter filling up the maximum area in a one by one mile section. These kinds of fields are common in Nebraska, where wells are drilled into the Ogallala Aquifer, and in Saudi Arabia, where wells are drilled into underground ancient river channels – sometimes more than a mile deep. Click on the [Photos] layer, search for the following lat/longs, and click on the photo icon to see what a center pivot system looks like. Then zoom out to about 35000 ft to see the circular fields.

40°33'02.90"N 101°49'09.11"W (Nebraska)
23°55'40.27"N 47°09'14.62"E (Saudi Arabia)

Traveling Through Stories

Lake Pepin and the remains of the Big Woods from Little House in the Big Woods. *The woods have been turned into farmland except along the streambeds. A debris dam from the Chippewa River created Lake Pepin, a part of the Mississippi River. (Google Earth Image, 44°25'38.94"N 92°05'52.55"W, eye alt 14 mi)*

Exploring the settings of books can bring new life and a new perspective to stories. Whether a children's book or a classic or the latest best seller, Google Earth always offers insight.

Most people have read at least one of the Little House books by Laura Ingalls Wilder, so let's begin by placemarking Pepin, Wisconsin near the site of the Little House in the Big Woods. Laura writes, "It was seven miles to town. The town was named Pepin, and it was on the shore of Lake Pepin". Zoom to an eye alt of about 20 mi. Use the Ruler to find the approximate location of Laura's house – anywhere in the hills north of Pepin. A replica of the house is located at 44°31'37N 92°11'25"W .

Laura wrote, "The great, dark trees of the Big Woods stood all around the house, and beyond them were other trees and beyond them were more trees. As far as a man could go to the north in a day, or a week, or a whole month, there was nothing but trees. There were no houses. There were no roads. There were no people. There were only trees and the wild animals who had their homes among them." Investigate the landscape in the region of Laura's house and think about how it has changed since Laura lived there; how the forest has mostly disappeared and been replaced by agricultural fields.

According to Laura, the town of Pepin is on the shore of Lake Pepin. Lake Pepin is a wide section of the Mississippi River. It was formed as debris from the Chippewa River created a dam in the Mississippi River (44°26'N 92°03'W; zoom out to 25 mi eye alt and look for the solid green region surrounding the Chippewa River which is the sediment dam). Pa said, "We can't get across the Mississippi after the ice breaks". Use the Ruler to measure how far they had to travel on the ice to cross the river.

Another book that lends itself to making Paths in Google Earth is Around the World in 80 Days by Jules Vern. Starting in London, England, placemark Phileas Fogg and Passepartout's main stops and draw their Path given the means of transportation that they used: London to Paris, France by rail and boat; Turin and Brindisi, Italy by rail; the Suez Canal and then Bombay, India by boat; Calcutta, India by rail; Hong Kong, China, Yokohama, Japan and San Francisco, California by boat; Salt Lake City, Utah, Chicago, Illinois and New York City by rail, and finally Liverpool, England by boat and London rail.

Here are some other books with interesting places to explore with Google Earth:

Make Way for Ducklings by Robert McCloskey tells the story of a family of ducks that make their way from a small island south of the Longfellow Bridge, Boston, Massachusetts to the Public Garden in the center of Boston, where the ducks find friendly swan boats.

The Little Red Lighthouse and the Great Gray Bridge by Hildegarde H. Swift tells of a small lighthouse under the east end of the George Washington Bridge. Zoom in to find the little lighthouse.

One Morning in Maine by Robert McCloskey tells the story of a girl named Sal on the day she lost her tooth. Guess which is Sal's island by finding Bucks Harbor, Maine. I think it is Bar Island. Many of Robert McCloskey's books are set in Maine.

Paddle to the Sea by Holling C. Holling lets you follow Paddle to the Sea's journey through Sault Ste. Marie, Lake St. Clair, over Niagara Falls, and along the St. Lawrence River to the Atlantic Ocean.

Dinosaurs of Waterhouse Hawkins by Barbara Kerley tells of some once-famous dinosaur statues. You can actually find those dinosaurs in Crystal Palace Park, London at 51°25'03.29"N 0°04'02.65"W.

Island of the Blue Dolphins by Scott O'Dell is a book we all read in 4th grade. Look for San Nicholas Island, California to see where Karana and Rontu became friends. The island is used as a weapons testing facility today – look for the air strip. Search for the Santa Barbara Mission, Santa Barbara to see where Karana and Rontu Aura were taken and where Karana soon died.

My Side of the Mountain by Jean Craighead George describes how a boy lived in a hollowed tree in the Catskills. Search for Delhi, New York where Sam visited the library, and look at the forests of the Catskills to see how hard it would be to live alone in a tree today and not be found.

Number the Stars by Lois Lowry tells of a girl who escapes the Occupation of Denmark during World War II. Search for Gilleleje, Denmark and use the Zoom and Look joystick to land on the shore and look across the water to Sweden to see what Annemarie saw when she was hiding in this remote village.

A Bear Called Paddington by Michael Bond is about a bear named after a train station. Search for London and search for Paddington Station, one of the grand train stations in London. Look for the train tracks into and out of the station, and then find Portobello Market. Use the Clock to find a day when the market is open.

Farewell to Manzanar by Jeanne Houston and James D. Houston is about a Japanese relocation camp in Manzanar, California. The buildings in Manzanar were mostly taken down after World War II, but a museum remains in the once-school auditorium and footprints of some of the buildings and the cemetery are still etched into the desert landscape.

From the Earth to the Moon by Jules Verne – Although this is a science fiction book written in 1865, it is interesting to search for some of the launch sites proposed in the book: Brownsville, Texas (being considered as a spaceport for some of the new commercial space launches), Longs Peak, Colorado, and Port Charlotte, Florida (across the state from Kennedy Space Center).

Sherlock Holmes by Sir Arthur Conan Doyle – Search for 221b Baker St., London, England. The place really does exist, although Sherlock Holmes did not.

The Adventures of Tom Sawyer by Mark Twain takes place in St. Petersburg, Missouri, which is really Hannibal, Missouri – the town where Mark Twain grew up. Just down the river is Fourmile Island – possibly Jackson Island where Tom and his friends camped and pretended they had drowned. If you also read The Adventures of Huckleberry Finn, you can track Huck and Jim's route down the Mississippi and see where they missed the Ohio River near Cairo.

Three Cups of Tea by Greg Mortenson and David Oliver Relin – Search for Korphe, Pakistan (probably possible because of the book's popularity) to find the small village where Greg Mortenson built his school. Can you find the school? Click on the [Photos] layer and look for the bridge, the fields, and the Braldu River. Search for K2 and check the elevation. Use the Look and Move joysticks to fly down the valley from K2 to Korphe. Look for the Baltoro Glacier (35°44'N 76°22'E) and think about how challenging it would be to climb up or down this rugged ice mass.

Da Vinci Code by Dan Brown – Follow Robert and Sophie through Paris, France (Ritz Paris and the Louvre Museum (use the [3D Buildings] layer to find the pyramid), Saint-Sulpic (48°51'04"N 2°20'05"E), and Chateau Villette, London (Westminster Abbey), and finally to the Rosslyn Chapel, Roslin, Midlothian, Scotland (there is a steel roof over the chapel in the image at the time of writing this book. It has since been removed – I hope you can see the chapel in a more up-to-date image).

Mapping Your Past

Dealey Plaza Park, Dallas, Texas is where President Kennedy was shot in 1963. The Texas School Book Depository is to the north of the plaza. Elm St., where the President's car was at the time of shooting, runs east/west in front of the Book Depository, and the grassy knoll is to the north of Elm St. (Google Earth Image 2012 Google, 32°46'44.63"N 96°48'31.88"W, eye alt 1600 ft)

After you have discovered where you live, you might consider looking up a few of the places your family has lived. Using Google Earth, you can create a kind of family tree using the folder system in the Places Panel. As an example, a family tree for President John F. Kennedy is presented below. As you can imagine, these family trees can get quite elaborate and can be organized in a variety of ways.

- 📁 John F. Kennedy
    - 📌 Born: 83 Beals Street, Brookline, Massachusetts
    - 📌 Summer home: 100 Marchant Ave., Hyannis Port, Massachusetts
    - 📌 Presidency: White House, Washington DC
    - 📌 Accomplishment: Kennedy Space Center, Florida
    - 📌 Died: Dealey Plaza Park, Dallas, Texas
    - 📁 Mother: Rose Fitzgerald Kennedy
    - 📁 Father: Joseph P. Kennedy, Sr.
        - 📌 U.S. Securities & Exchange Commission, Washington DC
        - 📁 Grandmother: Mary Augusta Hickey Kennedy
        - 📁 Grandfather: P. J. Kennedy
            - 📁 Great Grandfather: Patrick Kennedy
                - 📌 Dunganstown, County Wexford, Ireland
    - 📁 Brother: Robert F. Kennedy
        - 📌 United States Department of Justice, Washington DC
        - 📌 Died: Ambassador Hotel, 3400 Wilshire Boulevard, Los Angeles, California (demolished in 2005)
    - 📁 Brother: Ted Kennedy
        - 📌 Russell Senate Building, Washington DC
    - 📁 Sister: Eunice Kennedy Shriver
        - 📌 Soldier Field, Chicago, Illinois (site of first Special Olympics)
    - 📁 Wife: Jacqueline Bouvier Kenney Onassis
        - 📌 Married: St. Mary's Church, Newport, Rhode Island
        - 📌 Married (to Onassis): Skorpios, Greece
    - 📁 Daughter: Caroline Kennedy
    - 📁 Son: John Kennedy Jr.
        - 📌 Born: Georgetown University Hospital, Washington DC
        - 📌 Died: Atlantic Ocean off Martha's Vineyard, Massachusetts

Taking a Vacation

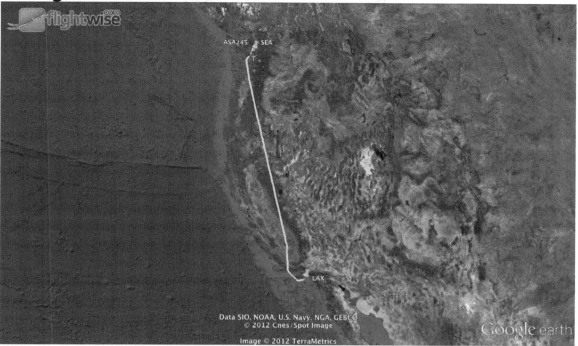

Flightwise track from Los Angeles to Seattle. Note that the flight passes near Yosemite and the Cascades. Mark down the name of each Cascade volcano and impress your fellow passengers. (Google Earth Image 2012 Cnes/Spot Image, Image 2012 TerraMetrics, Data SIO, NOAA, U.S. Navy, NGA, GEBCO, 40°N 122°W, eye alt 1500 mi)

Vacations past and present can be added as a set of Placemarks, a Path or a Tour. My favorite use of Google Earth for vacations is to plot where I am going – meaning the Path I will take in my car, plane or boat – to find out what I might see along the way. Plotting a Path for a car involves opening the [Roads] layer and using the highways to track your progress from home to your destination. You might also want to click on your personal folders to see if you are passing a relative or a place you visited in a book. Another layer that may be useful is the [US National Parks] layer under [More] → [Parks/Recreation Areas] in the Layers Panel. If you are traveling by boat, you can plot the ports of call with Placemarks and look for interesting coastlines.

If you are traveling by air, go to http://flightwise.com on the Internet to plot your airplane's flight track. On the Flightwise webpage, click on "Flight tracking" at the top of the page, and then "Track Flights" on the globe. Where it says, "Enter flight information", choose to search by "Airline". Then select your airline, type in your flight number, and click "Search". A new window will appear with a map of your flight line. Look to the right of the map for the "Google Earth" button and click it. Google Earth will open and your flight track will appear.

Use the navigation tools to investigate the flight track. Usually you can see ground features from your airplane window that are about the same distance from the line of flight as the altitude of the airplane. So at 30,000 ft, you can see features out to about 30,000 ft. A little trick is to fly along your flight track at about 10 miles eye alt – you should be able to see most of the features in the 3D Viewer on a clear day. You can find the flight track in the [Temporary Places] folder and drag it to a [Vacations] folder for future reference. Note that this track is an estimate and may change due to weather or air traffic.

Navigating the Nile River

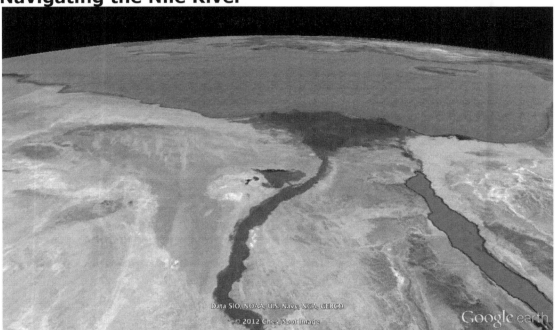

Perspective view of Egypt highlighting the Nile River, the floodplain, and the delta with the Mediterranean Sea in the background. (Google Earth Image Data SIO, NOAA, U.S. Navy, NGA, GEBCO, 2012 Cnes/Spot Image, 30°04'29.85"N 30°42'24.36"E, eye alt 130 mi)

Most astronauts come back from space with photographs of the Nile River because its dark green floodplain and delta create a stark contrast to the surrounding desert. To begin exploring the Nile Valley, return to the Great Pyramid of Giza and look for the three large pyramids of Khufu (northernmost pyramid), Khafre, and Menkaure (southernmost). The pyramids were built in the desert rather than on the floodplain because the dry desert air helped to preserve the bodies and belongings of the pharaohs. To the east of Khufu's pyramid are three smaller Queens' Pyramids, and to the east of these, and to the west of Khufu's pyramid are cemeteries. Perhaps the size of Khufu's pyramid relative to the size of his queens' provides some insight into the importance of the pharaohs relative to their queens.

Zoom out to an eye alt of about 15 mi and find the Nile River and the floodplain. Notice the sharp division between the green floodplain and the tan desert. Travel north to the Nile Delta and zoom in to see the patterns in the land created by the farmers who live in this region. Historically the Nile River flooded every late summer bringing water and nutrients to the floodplain and delta. But some years, too much water came down the Nile creating excessive flooding, and some years there was not enough water, creating drought conditions. The Aswan Dam now controls irrigation and flooding, and generates electricity. There are actually two dams, the Low Dam (24°02'02.01"N 32°51'56.77"E), which was completed in 1902, and the High Dam (23°58'14.68"N 32°52'39.01"E), built from 1960 to 1970.

Navigate to Alexandria, Egypt's largest port and once-location of one of the Seven Wonders of the World, the Alexandria Lighthouse. The lighthouse was destroyed over the years, and in its place is the Citadel of Qaitbay (search for Kaupay Fortress, Alexandria, Egypt), a fortress established in 1477 AD. Now fly west along the north coast of Egypt at an eye alt of about 20,000 ft and look for evidence of civilization. There are a number of resorts along this coast that take advantage of the sunny Mediterranean climate, aquamarine seas and white sandy beaches (e.g., 31°04'17"N 28°21'30"E). What do you think this region will look like in 20 years?

Roaming Around Rome

The Colosseum in Rome. (Google Earth Image, 2007, 41°53'24.96"N 12°29'32.33"E, eye alt 1400 ft)

When you walk through the streets of Rome, it is chilling to see the layers and layers of history built upon each other. Google Earth also provides a variety of layers to help you explore this powerful city. Search for Rome, Italy, and click open the [Rumsey Historical Maps] layer. Zoom in until two yellow compass roses show up. Choose the one labeled "Ancient Rome 1830" and open the historical map. Zoom out to see the full extent of the ancient map, and look for the Tiber River. Notice the island in the middle of the river. This island and Rome's distance inland allowed Rome to become very powerful. The island, and associated shallower parts of the Tiber, made Rome a center of trade, and Rome's distance inland from the Mediterranean offered protection from sea invasions.

Look for the hills of Rome in the Rumsey map. There are seven important ones. Notice the location of Circus Maximus and the Colosseum with respect to the hills. The Colosseum and Circus Maximum were built between the hills as a means of encouraging the independent settlements on the hills to interact. To appreciate the hills of Rome, click off the Rumsey map, placemark the southeast end of Circus Maximus and Trajan's Market (41°53'43"N 12°29'09"E), and draw a Path between the two. Navigate to the beginning of the Path in Circus Maximus and zoom to an eye alt of 500 m. Use the Move joystick to move along your Path.

Trajan's Market can be seen as it stands today by using Street View. Drag the Pegman over the market and drop it on Via Alessandrina where it turns blue. Use the North-up button to look around the market, which is built into the side of one of the hills.

The Roman Empire reached far beyond Rome. To appreciate its extent, placemark these sites:

Timgad, Algeria (35°29'N 6°28'E, 7000 ft eye alt)
Hadrian's Wall and Housesteads Fort, England (55°00'49"N 2°19'50"W, 2300 ft eye alt)
Pula Arna, Croatia
Pont du Gard, France - *aqueduct*
Palmyra, Syria (34°32'57"N 38°16'12"E, 4500 ft eye alt)
Segovia Aqueduct, Segovia, Spain (eye alt 4000 ft) – *Set the Clock slider to 8/2/2007 to appreciate what the Romans built by observing its shadow.*

Circling the Globe with Magellan

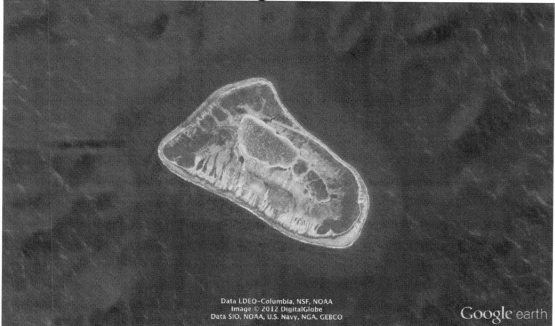

The island of Puka-Puka, one of the Tuamotus Islands in the middle of the Pacific Ocean, where Magellan landed. (Google Earth Image Data LDEO-Columbia, NSF, NOAA, Image 2012 DigitalGlobe, SIO, Navy, NGA, GEBCO, 14°49'08.56"S 138°48'36.14"W, eye alt 2900 ft)

Your appreciation of the wide range of scales in Google Earth will never be greater than when you follow the path of an explorer. Depending on the book you read about a particular explorer, the specific stops along his way may differ.

Magellan's crew was the first to circumnavigate the globe, although Magellan himself did not complete the voyage. His expedition sought the Spice Islands and, in his pursuit, Magellan was the first European to travel from the Atlantic to the Pacific Ocean, and the first to cross the Pacific Ocean. He began in Spain and died in the Philippines, but his men and ships made it back to Spain. Placemark these locations to track his voyage.

Seville, Spain – *Point of departure of Magellan's five ships. The ships sailed down the Guadalquivir River to San Lucar de Barrameda, Spain.*
Canary Islands – *The Canary Islands were a common stop for European explorers because of the favorable easterly trade winds.*
Rio de Janeiro, Brazil – *This was a known port at the time.*
Rio de la Plata, Uruguay – *This was possibly the first inlet that could have been the strait Magellan was seeking.*
Puerto San Julian, Argentina – *The second possible strait.*
Cape Virgenes, Argentina – *The beginning of the Strait of Magellan. Success! The strait is narrow and usually has stormy weather, so is hard to see with Google Earth. But zoom in and look hard and you will be able to follow Magellan's path from the Atlantic to the Pacific.*

Cabo Deseado, Chile (52°45'26"S 74°41'21"W) – *The end of the Strait of Magellan. Zoom down and use the Look joystick to look out over the Pacific and consider whether you would have set sail into this great ocean.*
Puka-Puka, Tuamoto Archipelago – *A tiny island in the middle of the Pacific where Magellan landed after not having seen land for many weeks. Note how far Magellan traveled between the Strait of Magellan and Puka-Puka. How in the world did Magellan ever find that tiny island?*
Pacific Ocean – *Zoom out to a global view that includes Tierra del Fuego to appreciate Magellan's voyage across the Pacific. How close to Antarctica did Magellan get?*
Guam – *Magellan sited three islands including Guam, Rota and Saipan and may have landed on any one of the three.*
Cebu, Philippines – *Magellan died on this island. How close to the Spice Islands, or the Moluccas, did Magellan get?*

Climbing High with Pizarro

Macchu Picchu in Peru – one of the Incan sites that Pizarro did not discover. (Google Earth Image 2012 DigitalGlobe, 13°09'49.58"S 72°32'42.21"W, eye alt 9700 ft)

Pizarro was a Spanish conquistador who is famous for conquering the Incas of Peru for their gold and silver. Although he never made it to Machu Picchu, he captured Atahualpa, the last emperor of the Incas; destroyed the Incan capital of Cusco; and founded Lima as the Ciudad de los Reyes. Important locations on Pizarro's journey are:

Isthmus of Panama – *This was Pizarro's staging ground.*
Puna, Ecuador – *He fought a battle here as he headed south.*
Tumbes, Peru – *Pizarro's initial goal; from here he headed inland.*

Paita, Peru – *Pizarro established this settlement as the first European city in Peru.*

Cajamarca, Peru (use lat long 07°09'52"S 78°30'38"W or Google Earth will fly you to Cajamarca Region, not city) – *Pizarro defeated Atahualpa here. Note the Yanacocha Gold Mine (use the Ruler to measure 12 miles north of Cajamarca). It is the second largest gold mine today. Perhaps some of the Inca's gold that lured Pizarro to Peru was mined here.*

Cuzco, Peru – *Inca capital conquered by Pizarro; check the elevation!*

Jauja, Peru – *Named by Pizarro as the first capital of Peru, but it was too far inland.*

Lima, Peru – *Final capital of Peru, also established by Pizarro. Look for the Cathedral of Lima and the Plaza Major (or Plaza de Armas) to find the center of the original city and original burial place of Pizarro.*

Machu Picchu is the most famous Inca site today. It was thought to have been built as an estate for the Inca emperor, Pachacut. Search for Machu Picchu and placemark it. Then fly back to Cuzco and use the Move joystick to fly up the Urubamba River Valley from Cuzco to Machu Picchu to appreciate how difficult it would have been for Pizarro to find this site.

Paddling Upstream with Lewis and Clark

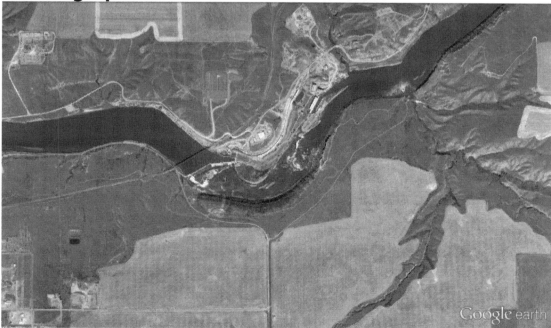

Crooked Falls, one of the five falls in Montana that caused Lewis and Clark to portage their gear and canoes for more than a month. (Google Earth Image, 47°32'04.18"N 111°11'53.96"W, eye alt 13000 ft)

With the Louisiana Purchase in hand, Lewis and Clark were commissioned by President Jefferson to find a water route from the east to the west. Rather than looking for a specific fort, town, or settlement along Lewis and Clark's trail, most of which are no longer there, follow the rivers and look for the challenges

Lewis and Clark faced. Remember that on the first part of the trip, they were traveling upstream. Here are some of the major landmarks:

Camp Dubois near Hartford, Illinois – *Staging area for Clark. He then headed up the Missouri River to Saint Charles, Missouri to meet up with Lewis.*
La Charrette (38°37'N 91°04'W) – *Last Euro-American settlement on the Missouri River.*
Fort Mandan – *Upriver a little from Washburn, North Dakota. Lewis and Clark wintered here and met Pocahontas. A replica of the fort is at 47°17'54.54"N 101°05'08.61"W.*
The Great Falls on the Missouri River in Montana – *The team had to portage around all of these falls, which took more than a month. The falls mark the limit of the westward-navigable portion of the Missouri River. The falls are Big Falls (47°34'07.22"N 111°07'24.98"W), Crooked Falls (47°32'10.03"N 111°11'47.57"W), Rainbow Falls (47°32'02.83"N 111°12'18.65"W), Colter Falls: (47°32'12.49"N 111°12'53.05"W), Black Eagle Falls (47°31'09.83"N 111°15'48.82"W). Note that several of the falls now have dams.*
Lemhi Pass (44°58'27"N, 113°26'42"W) – *This pass is in the Bitterroot Mountains, a part of the Rocky Mountains, and is on the Continental Divide. Watch the elevation as you navigate around Lemhi Pass to convince yourself that this is indeed the Continental Divide.*
Great Divide to the Pacific Ocean – *To follow the rivers to the Pacific, placemark Orofino, Idaho, five miles upstream from where Lewis and Clark put their canoes into the Clearwater River; Lewiston, Idaho, where they met the Snake River; and Kennewick, Washington, where the Snake River meets the Columbia River. Then follow the Columbia River to the Pacific.*
Fort Clatsop, Oregon – *At the mouth of the Columbia River and on the Pacific Ocean.*

Bringing World War II to America

USS Missouri (above water) and USS Arizona (below water) in Pearl Harbor, Hawaii. (Google Earth Image USGS, 21°21'47.60"N 157°57'07.40"W, eye alt 3000 ft)

America's entry into World War II came on December 7, 1941, the day Pearl Harbor was attacked. Search for Pearl Harbor, Hawaii and look around the bay for the USS Missouri (above water, 21°21'44.22"N 157°57'12.48"W) and the USS Arizona (under water, 21°21'53.86"N 157°56'59.92"W). Open http://www.gearthhacks.com and search for "Pearl Harbor V (Overlay), December 7, 1941". Look for the little Google Earth icon and click on it to download an overlay of an aerial view taken from a Japanese aircraft of Battleship Row after the USS Arizona was bombed.

Soon after the bombing at Pearl Harbor, internment camps for Japanese American citizens were established. Manzanar, California was one of these camps. Its remains can be seen in the Owens Valley close to Mount Whitney and Lone Pine, California. Search for Manzanar, California and look for the roads, a few trees, the footprints of a few buildings, and the Manzanar High School Auditorium, which has been turned into a museum.

From 1942 to 1945, the Manhattan Project developed the first atomic bomb at a secret location on a mesa in New Mexico called Los Alamos. The first bomb was detonated at the Trinity Site, New Mexico. Search for it, look for the circular fence, and then zoom out to get an idea of the remoteness of the test site. You can use the Overlay Tool to overlay a photograph of the site immediately following detonation (http://en.wikipedia.org/wiki/File:Trinity_crater_%28annotated%29_2.jpg). First use the Path Tool to make a line in the 3D Viewer that is 200 m in length, and then match that to the 200 m scale on the photograph. Rotate the overlay until the road patterns match. The second atomic bomb, called Little Boy, was loaded onto the Enola Gay airplane on Tinian Island in the Pacific. Search for Tinian Island and zoom out to observe its location relative to Japan. It was detonated over Hiroshima, Japan on August 6, 1945. Search for A-Bomb Dome, Hiroshima, Japan and enter street view to see the building closest to the hypocenter that remained standing. The third bomb, Fat Man, was detonated over Nagasaki on August 9, 1945. A circular park at 32°46'25"N 129°51'48"E commemorates the site.

Testing of nuclear bombs continued after the war with one of the most dramatic tests being that of the hydrogen bomb on Bikini Atoll, Marshall Islands on March 1, 1954. Search for Bikini Atoll and zoom out so the atoll fills the 3D Viewer. Look for the circular-shaped hole in the northwest part of the atoll; this is the Castle Bravo Site. Testing also continued at the Nevada Test Site – search for 37°06'N 116°03'W, eye alt 30,000 ft and look for the bomb craters. The Apollo astronauts trained at the Nevada Test site because of its many craters that are similar to those on the Moon, although created by very different processes.

Remembering the War in Europe

Auschwitz I concentration camps in Poland. (Google Earth Image 2012 GeoEye, 50°01'57.26"N 19°11'31.57"E, eye alt 9000 ft)

Most of the damage from World War II in Europe has been erased because buildings, bridges and airports have been reconstructed over the years. But historical imagery and remaining memorials can help us appreciate what happened. One of the locations that received extensive bombing was Warsaw,

Poland. Search for Warsaw, Poland and zoom in to the bridge at 52°14'10"N 21°02'28"E, eye alt 5000 ft. Use the Clock to look back to 1935 and then to 1945; the bridge was bombed along with the region to the northeast of the bridge. Look for the bomb craters at 52°14'20"N 21°02'42"E in the 1945 image, and then see what is there today.

One of the most unforgettable features of WWII was the concentration camp. Most camps have been destroyed, however, a few remain as memorials. One of the more well-known is Auschwitz in Poland. Auschwitz I is at 50°01'34"N 19°12'13"E and Birkenau (Auschwitz II) is at 50°02'16"N 19°10'35"E. The railroad tracks that brought in the prisoners are at 50°01'55"N 19°11'24"E between the two camps. Another was Dachau, about 10 miles northwest of Munich, Germany. The barracks are at 48°16'10"N 11°28'05"E and the crematorium area stands at 48°16'18"N 11°27'53"E.

The invasion of Normandy started along the southern coast of England, near the Isle of Wight, crossed the English Channel, and landed on Omaha (49°22'37"N 0°53'25"W), Utah (49°24'42"N 1°10'10"W), and Gold (49°20'40"N 0°34'50"W) beaches, among others. Today the beaches border the peaceful farmlands of northern France. Note the oyster and mussel farms along Gold Beach.

Finishing the War in the Pacific

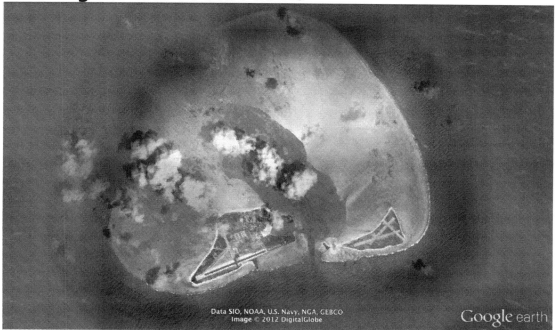

Midway Atoll in the Pacific Ocean, site of the Battle of Midway. (Google Earth Image SIO, NOAA, U.S. Navy, NGA, GEBCO, 2012 DigitalGlobe, 28°13'56.07"N 177°21'57.17"W, eye alt 43000 ft)

Guadalcanal, Saipan, and Iwo Jima bring images of bloody conflicts in the Pacific. Below are some of the more famous battles. Plot these islands on

Google Earth using the Placemark tool to see how the battles moved across the Pacific and approached Japan. Look at the islands and decide if they are atolls or islands. Look for military installations and evidence of people living on the islands today.

Battle of Wake Island – *December 7-23, 1941*
Battle of Midway (28°12'N 177°21'W) – *June 4-6, 1942*
Battle of Guadalcanal – *August 7, 1942 - February 9, 1943*
Battle of Tarawa (Tarawa Atoll) – *November 20, 1943*
Battle of Kwajalein (Kwajalein Atoll, Marshall Islands) – *February 1, 1944*
Battle of Saipan – *June 15, 1944*
Battle of Guam – *July 21, 1944*
Battle of Tinian – *July 24, 1944*
Battle of Peleliu (present day Palau) – *September 15, 1944*
Battle of Iwo Jima (Iwo Jima, Japan) – *February 19, 1945; the famous flag-raising was on top of Mount Suribachi.*
Battle of Okinawa - *April 1, 1945*

Following Current Events

One of the tent cities in Port-au-Prince. This one is just to the east of the National Palace, which was badly damaged in the 2010 earthquake. (Google Earth Image, 18°32'32.19"N 72°20'12.67"W, eye alt 1300 ft)

More and more TV news stations are using Google Earth images to orient viewers to the locations of news stories. Google Earth can bring you right into the story and, using the Clock, sometimes allow you to follow the story over time.

In 2010, a strong 7.0 earthquake struck Port-au-Price, the capital of Haiti. The National Palace (18°32'35"N 72°20'20"W) was badly damaged and thousands of people were left homeless. Tents were set up in parks around the city as temporary homes. Many of these tent cities remain at the time of writing this book in 2012. Search for the National Palace, Port-au-Prince, Haiti and look at the palace over time using the Clock. The earthquake happened on January 12, 2010. Look to the east of the National Palace to find one of the tent cities and use the Clock to find out when it was set up and whether it still exists.

Another earthquake struck near Japan on March 11, 2011. The earthquake was followed by a tsunami. The resulting damage to the Fukushima I Nuclear Power Plant (37°25'16"N 141°01'57"E) is apparent in the Google Earth image. Reactor 4 (37°25'11.86"N) had been defueled., but reactors 1 (37°25'23.10"N), 2 (37°25'20.04"N) and 3 (37°25'16.11"N) experienced meltdowns after the earthquake and tsunami.

On January 20, 2009, Barack Obama was inaugurated as the 44th President of the United States on the steps of the Capital in Washington DC. Search for Washington DC and navigate to a view that includes the Capital and the Mall. Use the Clock to find January 2009. The blobs of "ants" that populate the mall are people who have come to watch the inauguration.

On September 11, 2001, the World Trade Center towers were hit with airplanes in an act of terrorism. Search for the World Trade Center, New York City, New York and look for the new construction, including the two square ponds that mark the sites of the towers. Use the Clock to go back in time before and just after September 11, 2001 to appreciate the magnitude of this disaster.

On May 2, 2011, the mastermind of the World Trade Center bombings, Osama Bin Laden, was killed in Pakistan. Search for "Osama Bin Laden's Hideout Compound, Pakistan" (34°10'09.57"N 73°14'32.90"E). You may need to use the Clock to find it as it has been torn down.

Finally, the Costa Concordia hit a rock off Isola del Giglio in the Tyrrhenian Sea on January 13, 2012 (42°21'55"N 10°55'17"E). The cruise ship has been imaged and can be seen on Google Earth. You may need to use the Clock to find it as it has been towed away. You can also see it by clicking on the [3D Buildings] layer. How close was the ship to the shore and surrounding rocks?

Surviving Extreme Weather

Track of the tornado that struck Joplin, Missouri in 2011. (Google Earth Image USDA Farm Service Agency, 2012 GeoEye, 37°03'49.10"N 94°30'01.48"W, eye alt 35000 ft)

Tornadoes, hurricanes, and floods strike on very different space and time scales, but you can see their effects in Google Earth.

Tornadoes strike relatively small regions for a short period of time, so seeing a tornado in a Google Earth image is nearly impossible. Also, tornadoes are hidden from above by a layer of clouds. It is, however, possible to map the track of a tornado because of the path of destruction left behind. Search for Joplin, Missouri; a tornado struck this area on May 22, 2011 and created a path of destruction from 37°03'29"N 94°33'20"W to 37°04'19"N 94°28'06"W and beyond. Look for the light brown diagonal strip across the city. Zoom in to the light brown region to see the destruction and move north or south to see what the area looked like before the tornado struck. If you don't see it, you may need to click on the Clock and look back in time as houses have been rebuilt after the tornado. Look for other paths of tornados in Tuscaloosa, Alabama (April 27, 2011 – use the Clock) and White Lake, Wisconsin. You will need to zoom out to about 18 mi eye alt and may need to use the Clock.

Hurricanes occur at a much slower rate than tornadoes and can be tracked using the [Weather] layer in the Layers Panel. You cannot, however, go back in time to see the hurricane because the [Weather] layer has only current weather data. If you hear about a hurricane that is in the process of building, click on the [Clouds] layer in the [Weather] layer and navigate to a regional view (eye alt of 500 - 1000 mi) of the area in which the hurricane has been reported. Create a new folder named after the hurricane and add a Placemark placed at the eye of the hurricane. Name the Placemark with the current date. Tomorrow add another Placemark, and continue on as the hurricane progresses up the coast of the United States or into the Gulf of Mexico.

The Clock is very handy for tracking floods because you can obtain a before and after picture once you know a flood has occurred. Navigate to Memphis, Tennessee and use the Clock to find the flood of the Mississippi River, which occurred April to May, 2011.

More flooding:
Ban Pom, Phra Nakhon Si Ayutthaya, Thailand (Chao Phraya River) – *November 2011*
Fort Calhoun Nuclear Power Plant, Blair, Nebraska (Missouri River) – *Summer 2011*
Cairo, Illinois (Mississippi River) – *May 2011*
New Orleans, Louisiana (30°01'3"N 90°07'18"W and 29°59'34"N 90°05'14"W, eye alt 1300 ft) – *Use the Clock to find 7/12/2005 and then 8/31/2005. Zoom out to see which areas were affected by the breach of the levees.*

If you add a flood or tornado to your folders, it is a good idea to include the date of the event in the name of the Placemark.

Watching Volcanoes Erupt

Mt. Saint Helens, Washington after it erupted. (Google Earth Image 2012 DigitalGlobe, 2012 GeoEye, USDA Farm Service Agency, 46°14'13.92"N 122°8'45.47"W eye alt 16 mi)

Volcanoes are most exciting when they are erupting, but catching eruptions in satellite images is challenging. The astronauts on the Shuttle and Space Station have had good luck photographing eruptions (see http://earthobservatory.nasa.gov and search for "erupting volcano"). But seeing the after-effects of volcanic eruptions on Google Earth can help us appreciate the potential devastation of these eruptions.

Look in the Layers Panel under [Global Awareness] for a layer called [UNEP: Atlas of Our Changing Environment] and click it open. Look for the blue squares, which indicate places where change has occurred on Earth for a variety of reasons. Find the [UNEP: Atlas of Our Changing Environment] in the [Temporary Places] folder and click it open. Search for Mt. Saint Helens and click on the blue square. In the popup window, click on "Overlay Images in Google Earth". The most recent of three images of Mt. Saint Helens will appear. Look in the [Temporary Places] folder for the [Mount St. Helens, United States] folder. Use your cursor to successively click each image on and off using the box to the left of the Placemark for each date. Find the mudflow zone (area where volcanic mud flowed down the mountain), the blast zone (area where the blast from the eruption blew over trees), and the post-eruption crater. The blast zone can be identified because new vegetation would be appearing in the 07 September 1999 image, while there would be little new vegetation where the mudflows occurred. During the 1980 eruption, the first thing to occur was the collapse of the north flank of the volcano. Since the eruption, a dome has formed in the crater left by the eruption. Use the Zoom and Look joystick to investigate the collapsed north flank and the dome.

Mt. Vesuvius is another famous volcano in Italy partly because the city of Pompeii was buried during the AD 79 eruptions. Search for Mt. Vesuvius, Italy to an eye alt of 30,000 ft. Look for the crater and evidence of past eruptions in the form of ridges radiating from the center of the volcano. Navigate to the southeast of the mountain to a latitude and longitude of 40°45'00"N 14°29'10"E and an eye alt of 2000 ft. This is the city of Pompeii that was buried under 15 feet of ash. Much of the city has been excavated starting in 1599. The houses may look similar to those in your neighborhood from this perspective, but they have no roofs! Look for the amphitheater (southeast of city: 40°45'4.53"N 14°29'42.45"E), the theater (40°44'54.38"N 14°29'19.25"E) and the forum (southwest of city: 40°44' 57.78" N, 14°29' 4.90"E).

Discovering Impact Craters

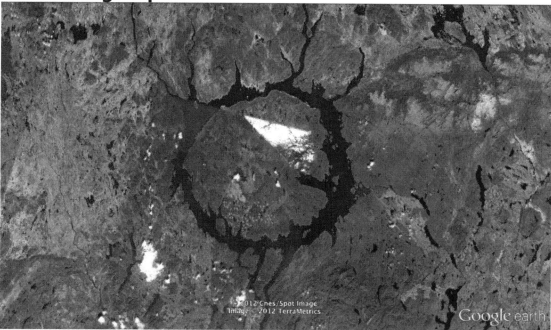

Manicouagan impact crater in Canada. (Google Earth Image 2012 Cnes/Spot Image, 2012 TerraMetrics, 51°23'25.70"N 68°37'55.51"W, eye alt 75 mi)

When you look at the Moon, impact craters are hard to miss, especially on the far side of the Moon. Although these craters were formed many millions of years ago, the Moon has no water or atmosphere to erode the craters, so they remain as they were when they were formed. Earth also was heavily impacted by meteors millions of years ago, but water and wind have erased most of the resulting impact craters. There are exceptions, however, the largest being Manicouagan in Canada. Search for Lac Manicouagan, Quebec, Canada at an eye alt of about 100 mi and look for the circular lake that highlights the ring of the crater.

Other impact craters:
Meteor Crater, Arizona
Bosumtwi, Ghana
Aorounga, Chad
Gosses Bluff, Australia

One of the largest known impact craters on Earth is called Chicxulub. This impact may have been responsible for the extinction of the dinosaurs 65 million years ago. An asteroid hit Earth near the Yucatan Peninsula in Mexico and created a crater 110 to 300 mi in diameter. Evidence of the crater had been challenging to find, but one group of scientists traced the crater by looking for cenotes, or sinkholes with rocky edges that are filled with water, which formed along the crater's rim. Below are the lat/longs of five of these cenotes. You will need to zoom in to an eye alt of about 20,000 ft to see them. Placemark the

cenotes, and look for others and placemark them. Use the Path tool to trace out the crater. Where is its center? What is the diameter?
    Cenote #1 20°39'22.62"N 89°12'03.30"W
    Cenote #2 20°35'06.80"N 89°29'16.13"W
    Cenote #3 20°36'01.21"N 89°41'43.97"W
    Cenote #4 20°46'42.32"N 90°02'09.95"W
    Cenote #5 21°10'51.27"N 88°43'34.39"W

Digging for Energy

Antelope Coalmine Road, Douglas, Wyoming. Note the circular portion of the railroad track where the cars are loaded with coal. (Google Earth Image 2012 DigitalGlobe, 43°28'6.77"N 105°20'27.40"W, eye alt 17000 ft)

The majority of the world's energy to run our factories, heat our homes, and drive our cars comes from fossil fuels buried in the ground. The coalmines and coal fired power plant, and the oil and gas wells and refineries can be easily seen on Google Earth.

<u>Coalmines</u>
Antelope Coalmine Road, Douglas, Wyoming – *Zoom in on the loop in the railroad tracks and look for the whitish cars that are empty, and the black cars that are full. Travel north along the tracks and look for more mines at each loop, and travel south and see how many trains you can find on the tracks.*
Wuhai City, Inner Mongolia, China (39°20'N 106°55'E)
Kusmunda, India – *Search for Gevra, India*
Also look in the [Appalachian Mountaintop Removal] layer in the [Global Awareness] folder in the Layers Panel.

Coal Fired Power Plants
Drax, Britain (53°44'14"N 0°59'52"W, eye alt 6,000 ft) – *Notice the steam coming from the plant. This is water – a byproduct of combustion – not carbon dioxide. The carbon dioxide is there, but it is invisible.*
Civaus, France (46°27'19"N 0°39'19"E)
Kingston, Tennessee (35°53'53"N 84°31'17"W)

Oil and Gas Wells (usually found together):
Ghawar Fields, Saudi Arabia (24°49'0"N 49°14'33"E, eye alt 7500 ft)
Denver City, Texas – *Zoom to 10 mi to start to appreciate the fields.*
Prudhoe Bay, Alaska – *Zoom to 30,000 ft and look south of the bay*
Alaska Pipeline from Prudhoe Bay to Valdez, Alaska – *Zoom in to 68°58'59"N 148°49'45"W to see a part of it.*
Cabimas, Venezuela – *Search for 10°17'27"N 71°19'19"W and zoom to an eye alt of 20,000 ft.*
Athabasca Tar Sands, Canada – *Search for Fort McMurray, Canada and zoom to 90 mi; then look north of Fort McMurray for the tar sands.*
Oil platforms off Santa Barbara, California (34°19'57"N 119°37'21"W)
Kuwait Oil Fields, Kuwait (28°54'0"N 47°57'50"E)

Oil Refineries
Standard Oil, El Segundo, California (33°54'33"N 118°24'44"W)
Marathon Texas City Refinery, Texas City, Texas
Reliance Refinery Complex, Gujarat, India

Deforesting Our Planet

Clearcutting near Rondonia, Brazil in 2012. (Google Earth Image 2012 Cnes/Spot Image, 10°31'26.18"S 64°1'50.66"W eye alt 100 mi)

Carbon stored in forests, especially the tropical forests of the Amazon, is contributing to climate change because the cut forests are usually burned, thus releasing carbon into the atmosphere as carbon dioxide. Another way to think about this is the conversion of forested land to crops and grazing land significantly reduces the amount of carbon that is stored on a given section of land. To appreciate the extent of the deforestation, use the Clock to investigate these locations:

Rondonia, Brazil – *Look everywhere to see extensive clearcutting. Zoom in to 9°52'55"S 63°45'01"W, eye alt 100 mi, and use the Clock to go back in time to watch the clearcutting (in reverse). Note the unusual clearcutting patterns at 19°57'S 59°18'W.*

Borneo – *Much of the native forest is being replaced with palm oil plantations. Palm oil is used for cooking, as an engine lubricant, in cosmetics, and as a biofuel. These plantations take away a diverse tropical forest that is home to orangutans. Search for 5°34'60"N 118°4'0"E and zoom in to an eye alt of 700 ft to see the palm trees; then zoom out and fly around to appreciate the extent of these plantations.*

Washington – *Search for Mt. Saint Helens, Washington and look about 20 miles to the west for one of many examples of deforestation in the United States. Use the Clock to see how rapidly this land changes; including by reforestation.*

Renewing Our Energy

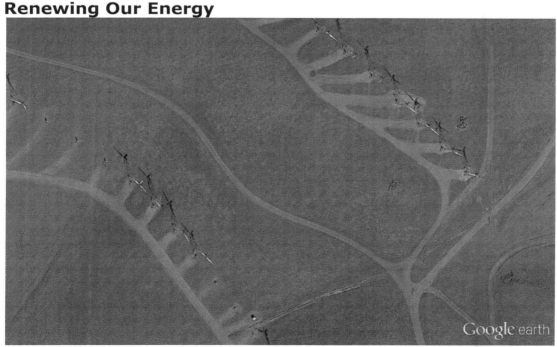

Windmills near Livermore, California. (Google Earth Image, 37°43'4.97"N 121°38'34.70"W, eye alt 2300 ft)

There are many ideas for converting to carbon free sources of energy, including windmills, solar farms, solar panels on our roofs, geothermal plants, and tides. In some areas, large wind and solar farms have already sprung up, and solar panels on rooftops are more and more apparent in Google Earth. And of course hydroelectric power from dams has been a carbon-free source of energy since ancient times.

Wind Farms
Altamont Pass, California (37°43'N 121°39'W)
San Gorgonio Pass, California (33°56'39"N 116°39'56"W)
Sweetwater, Texas (32°19'43"N 100°21'29"W)
Ocean windmills off Denmark (55°41'14"N 12°40'11"E)

Solar Farms
Daggett, California – *Navigate to about 3 mi east of Daggett, California in the Mojave Desert. Solar One is at 34°52'19N 116°50'3"W and uses a system in which mirrors are pointed at a central tower where electricity is generated. The solar plant to its southeast (34°52'6"N 116°49'32"W) uses a different system in which energy is collected directly on the solar panels, which are aligned in rows.*
Kramer Junction, Mojave Desert, California (35°0'54"N 117°33'33"W)
Olmedilla Photovoltaic Park, Spain (39°37'19"N 2°5'4"W)
Waldpolenz Solar Park, Germany (51°19'42"N 12°39'22"E)
Moura Photovoltaic Power Station, Portugal (38°11'37"N 7°12'58"W)

Solar Panels on Rooftops
Google Headquarters, Mountain View, California
Bollinger Canyon Elementary School, San Ramon, California

Hydroelectric
Hoover Dam – *Across the Colorado River on the Arizona/Nevada border*
Three Gorges Dam, China – *Across the Yangtze River. Search for Sandouping, China.*
Wilson Dam, Alabama – *On the Tennessee River. The first of nine dams built for the Tennessee Valley Authority.*

Geothermal Plants
Mammoth Pacific Geothermal Facilities – *Search for Substation Rd., Mammoth Lakes, California. This plant was built in a geothermal area related to the Long Valley Caldera.*
Reykjanes Power Station, Iceland (63°49'33"N 22°40'58"W) – *Use the Clock to find a day without clouds.*

Painting Earth

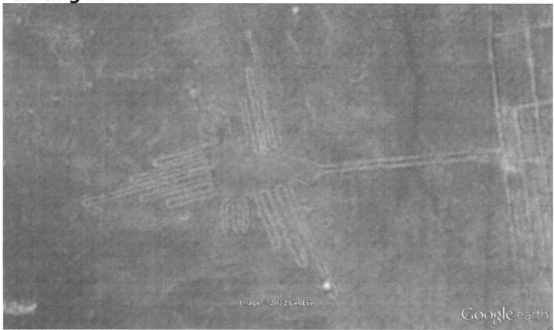

The Colibri, or the Hummingbird, is one of the many Nazca Lines in Peru. (Google Earth Image 2012 GeoEye, 14°41'31.62"S 75°8'55.95"W, eye alt 1800 ft)

My students thought that center pivot irrigation fields provided decorations for the astronauts. What did the Nazca people think when they drew their gigantic drawings on the deserts of Peru? The Nazca Lines were made between 400 and 650 AD on a high desert plateau. Because rainfall is rare, the lines remain today. By removing the overlying reddish iron oxide-coated rocks, the underlying creamy-pink soil was exposed to create figures observable from above. Use an eye alt of 1800 ft to see the figures.

Manos (14°41'40"S 75°6'50"W)
Spiral (14°41'18"S 75°7'22"W)
Condor (14°41'51"S 75°7'34"W)
Colibri = Hummingbird (14°41'32"S 75°8'56"W)
Astronauta = Astronaut (14°44'43"S 75°4'47"W)

In more recent times, especially with more satellite images available to the public, decorating Earth with art, advertisements, and markings has become more popular. You can search the Internet and find many of these. Some of the most popular are listed below. These are usually best seen from low altitudes of about 2,000 ft or less and often are short-lived so you might need the Clock to find them.

Crop circles in England (52°43'8"N 1°38'49"E, Clock: 9/11/06)
Man-shaped lake in Brazil (21°48'19"S 49°05'24"W, Clock: 6/15/2011)
Opra, near Gilbert, Arizona (33°13'33"N 111°35'49"W, Clock: 12/31/2012)
Coca Cola logo in Chile (18°31'45"S 70°15'00"W, Clock: 5/11/2011)

Firefox logo in Oregon (45°07'26"N 123°06'50"W, Clock: 8/13/2006)
Del Coronado Hotel sand dunes in San Diego, California (32°40'57"N 117°10'56"W)
Target symbol near Chicago, Illinois (42°00'24"N 87°53'14"W)
Big Brother Eye in Queensland, Australia (27°51'42"S 153°18'51"E)
BMW in Munich, Germany (48°10'35"N 11°33'33"E)
Dole Plantation Maze in Hawaii (21°31'30"N 158°02'15"W)
Colonel Sanders in Nevada (37°38'46"N 115°45'03"W, Clock: 11/14/2006)
Thumbprint in a park in England (50°50'39"N 0°10'20"W, Clock: 4/15/2007)
Horse in the grass in Wales (51°39'03"N 3°15'22"W, Clock: 12/2001-12/2010)
– *See the horseshoes?*
Pink rabbit in Italy (44°14'39"N 7°46'11"E, Clock: 12/29/2012)
World logos in Washington (48°1'43"N 122°9'52"W, Clock: 12/2003) – *There is another logo on 8/2006!*

Beyond the decorations that people add to Earth, Mother Nature also has had a hand in painting our planet. Search for some of my favorite places that could be framed as art. Perhaps you will discover your own art gallery!

Crops in Brazil (15°26'51"S, 53°41'35"W, eye alt 50,000 ft)
Wetlands in Bolivia (12°53'S 65°44'W, eye alt 35 mi)
The Andes, Chile (33°39'S 70°04'W, eye alt 36,000 ft)
Glaciers in Chile (48°57'S 73°49'W, eye alt 10 mi)
Alluvial fans in Mongolia (42°31'N 105°31'E, eye alt 30,000 ft)
The Ganges Delta, India (21°45'N 88°22'E, eye alt 40 mi)
Lake Carnegie, Australia (26°12'S 122°40'E, eye alt 50 mi)
Roden Crater, Arizona (35°25'N 111°15'W, eye alt 11,000 ft)

Or look in the [The Earth from Above with GoodPlanet] layer under [Global Awareness] in the Layers Panel. If you click one of the green icons, you will see a photograph of an impressive feature. If instead you zoom in to where the green icon is located, you will likely see Google Earth's view of this special feature.

Investigating Religious Centers

The Vatican. (Google Earth Image, 41°54'7.50"N 12°27'23.22"E, eye alt 1700 ft)

Not only are there many religious centers in the world, but also many are important to more than one religion. Here are a few to get you thinking.

Christianity
Cathedral Santiago de Compostela, Spain – *Four major pilgrimage routes lead to this cathedral.*
Basilica of Our Lady of Guadalupe, Mexico City, Mexico – *The most visited Catholic church and beautiful from above.*
Notre Dame, Paris, France – *Use the [3D Buildings] layer to appreciate the flying buttress architecture.*

Important places in the life of Jesus include the following. Look at all the Placemarks from eye alts of above 50 mi to appreciate where Jesus spent his life.
Bethlehem, Israel
Nazareth, Israel – *Mary and Josef left Nazareth near the Sea of Galilee and traveled to Bethlehem where Jesus was born. Note that the first part of the journey is over relatively flat land, but the latter part crosses hills and valleys for many miles. Use the Directions tab in the Search Panel to find one possible route, and use the Move joystick to follow their route through the hills.*
Jerusalem, Israel – *Jesus lived in Nazareth and died in Jerusalem. The Church of the Holy Sepulchre, Jerusalem, Israel is the site of the crucifixion.*
Sea of Galilee, Israel

The Vatican City – *The Vatican is a walled city-state within the city of Rome but independent of Italy. Click on the [Borders and Labels] layer to see its boundaries. Look for St. Peter's Square, the dome of Saint Peter's Basilica, and Castle St. Angelo (especially if you are reading The Da Vinci Code by Dan Brown). The Sistine Chapel is in the Apostolic Palace, the residence of the Pope, at 41°54'11"N 12°27'16"E.*

Salt Lake Temple, Salt Lake City, Utah – *Use the [3D Buildings] layer to find it. Also look for the Mormon Tabernacle.*
Mormon Trail – *Starting in Nauvoo, Illinois, cross the Mississippi River to Council Bluffs, Iowa; cross the Missouri River and continue west along the north shore of North Platte River to Fort Kearney, Nebraska and Confluence Point – the confluence of the North and South Platte (present day North Platt, Nebraska). Continue on to Scotts Bluff, Nebraska (use the Zoom and Look joystick to fly around the bluff); Fort Laramie, Wyoming (42°12'11"N 104°33'26"W to see the fort); Independence Rock, Wyoming; Fort Bridger, Wyoming; over the Wasatch Mountains; and into the Salt Lake Valley through Immigration Canyon (it ends at the present day University of Utah).*

Mount Ararat, Turkey – *Zoom in to see if you can find evidence of Noah's ark.*

California Missions – *Mission San Francisco de Asis (37°45'51"N 122°25'37"W) is one of the prettiest from above, while Mission San Carlo Borromeo de Carmel (36°32'33"N 121°55'11"W) has a more characteristic shape with the central courtyard.*

Islam
The Quran states that the Kaaba was built by Abraham and Ishmael. All Muslims face the Kaaba to pray no matter where they are.

Masjid al-Haram, or Grand Mosque, Mecca – *It surrounds the Kaaba and is the site of the Hajj, an annual pilgrimage to Mecca. The Grand Mosque is the largest mosque in the world. Use the Clock to find dates when people are visiting.*
Mina, Mecca, Saudi Arabia (21°24'48"N 39°53'46"E) – *Hajj tents*
Al-Masjid al-Nabawi, or Prophet's Mosque, Medina – *Built by Muhammad and is Muhammad's final resting place.*
Al-Aqsa Mosque, Jerusalem, Israel – *The third holiest site in Islam and is located in the Old City of Jerusalem.*
Blue Mosque, Istanbul, Turkey (41°0'19"N 28°58'37"E) – *Built near the Hagia Sophia. Istanbul was originally called Constantinople and played many roles in history. Zoom out to an eye alt of 2600 mi to appreciate why the location of this city was so important in the history of Europe and Asia.*

Hinduism
Hindu is a religious, cultural and philosophical belief that originated in India.

Mount Kailas, Tibet – *The mountain is the home of Lord Shiva, a major Hindu deity, and a place of eternal bliss. The glaciers on this mountain are sources of the Indus, the Brahmaputra, and a tributary of the Ganges Rivers. Use the Zoom and Look joystick to see the unique structure of this sacred mountain. It is also a sacred mountain for three other religions: Bon, Buddhism, and Jainism.*

The Char Dham are four holy sites for Hindu pilgrimages in India:
Jagannath Temple, Puri, India
Badrinath Temple, Badrinath, India – *A low resolution image but the valley is breathtaking*
Dwarakadheesh Temple, Dwarka, India (22°14'16"N 68°58'03"E) – *This temple is dedicated to Lord Krishna.*
Ramanatha Swami Temple, Rameshwaram, India

Other important temples include:
Golden Temple of Amritsar, Punja, India
Swaminarayan Akshardham Temple, Delhi, India

Buddhism
Buddhism is a religious and philosophical belief based on the teachings of Buddha who lived and taught in the eastern part of India. Four important cities in the life of Buddha are Lumbini, Bodhgaya, Sarnath and Kushinagar. Placemark them to see how far Buddha traveled in his lifetime.

Lumbini, Nepal – *Buddha's birth place.*
Bodhgaya, India – *The place of Buddha's Enlightenment – the Bodhi Tree is at 24°41'45.35"N 84°59'28.49"E.*
Dhamek Stupa, Sarnath, India – *Buddha delivered his first teaching at Dhamek Stupa.*
Kushinagar, India – *Where Buddha died.*

Other important sites are:
Great Stupa, Sanchi, India (23°28'50.33"N 77°44'10.69"E) – *The oldest stone structure in India from the 3rd century BC.*
Mount Tai, China – *One of the five sacred mountains; under a cloud at the time of writing this book.*
Shwedagon Pagoda, Yangon, Burma (16°47'54"N 96°8'59"E) – *The most sacred Buddhist pagoda for the Burmese.*
Potala, Lhasa, Tibet – *The Potala was the main residence of the Dalai Lama until the 1959 Tibetan uprising. Zoom out to see the isolated location of the palace. The Potala is highlighted in a children's book called <u>Tibet Through the Bed Box</u> by Peter Sis. The Kyi River, known to the Tibetans as the merry blue waves, provides water for this isolated city; its sources are glaciers in the Himalayas. The Kyi feeds into the Yarlung Zangbo River; the original source of the Yarlung Zangbo River is Mount Kailas, Tibet!*

Judaism
Judaism is a religion and philosophy that is based on the laws and commandments of Moses as revealed on Mount Sinai.

Mount Sinai, Egypt – *Mount Sinai is on the Sinai Peninsula in Egypt. Search for St. Catherine's Monastery, Egypt (28°33'20"N 33°58'33"E); Mount Sinai is to the west of the monastery. Use the Move joystick to fly over Mount Sinai to appreciate its ruggedness.*
Western Wall, Jerusalem, Israel (31°46'36.45"N 35°14'4.44"E) – *Also known as the Wailing Wall, the resolution is low but look back in time using the Clock to find the shadow of the wall.*
Temple Mount, Jerusalem, Israel (31°46'40.83"N 35°14'7.29"E)

Celebrating Holidays

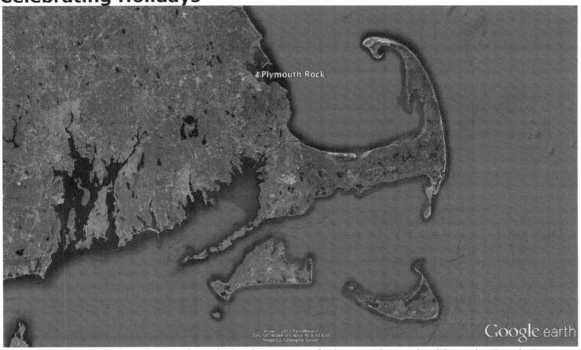
A regional view of Cape Cod and the location of Plymouth Rock. (Google Earth Image SIO, NOAA, U.S. Navy, NGA, GEBCO, U.S. Geological Survey, 2012 TerraMetrics, 41°42'N 70°28'W, eye alt 75 mi)

Holidays offer an interesting cross section of Google Earth imagery. The holidays presented here are those celebrated primarily in America.

New Year's Day: *The Rose Parade along Colorado Blvd., Pasadena, California between Orange Grove Blvd. and Sierra Madre Blvd., followed by the Rose Bowl game in the Rose Bowl Stadium, Pasadena, California.*

Martin Luther King's Birthday: *The Lincoln Memorial overlooking the Reflection Pond on the National Mall, Washington DC is where Martin Luther King gave his famous "I Have Dream" speech.*

Super Bowl Sunday: *The Super Bowl moves from stadium to stadium. In 2011 it was in the Cowboys Stadium, Arlington, Texas; in 2012 it was in the Lucas Oil Stadium, Indianapolis, Indiana; and in 2013 it was in the Louisiana/Mercedes-Benz Superdome, New Orleans, Louisiana. Use the Clock to see the Superdome soon after Hurricane Katrina struck in August 2005.*

The Presidents' Birthdays: *It is difficult to find interesting birthplaces of presidents, but each president is associated with certain historical sites. Mount Vernon, Virginia (search for "George Washington's Mt. Vernon Estate") is where George Washington lived, and Gettysburg National Military Park, Pennsylvania is where Abraham Lincoln gave his famous Gettysburg Address.*

Saint Patrick's Day: *The Chicago River in Chicago, Illinois is dyed green every Saint Patrick's Day.*

Earth Day: *Anywhere you like, or the global view.*

Independence Day: *Independence Hall, Philadelphia, Pennsylvania is where the Declaration of Independence was signed.*

Memorial Day: *Arlington Cemetery in Arlington, Virginia is one of many cemeteries where you might celebrate this holiday.*

Columbus Day: *Palos de la Frontera, Spain is where Columbus departed, and The Bahamas is where he first landed, although the specific island is not known. Columbus also visited Cuba, Haiti and the Dominican Republic before heading back to Spain on his first voyage.*

Halloween: *Perhaps the world's largest corn maze is near Dixon, California (38°29'00"N 121°49'10"W). Be sure you are looking at an image date in October. Anyone seen a sincere pumpkin patch?*

Veterans Day: *The Vietnam Veterans' Memorial and the Korean War Memorial or World War II Memorial (38°53'22"N 77°2'27"W) in Washington DC are certainly good places to honor our veterans.*

Thanksgiving: *Search for Plymouth, Massachusetts and look for Plymouth Rock (41°57'29.05"N 70°39'43.37"W); the rock is in a cage! Also look for the Mayflower II (41°57'35"N 70°39'43"W, unless it is sailing) and the Plimouth Plantation – a replica of the original plantation. You might also consider following the Macy's Thanksgiving Day Parade in New York City. Start at 77th St. and Central Park West, head south to Columbus Circle, and then east along Central Park South. Make a right turn at 7th Ave. and go south to Times Square. Turn left at 42nd St., right at 6th Ave., and right at 34th St. (Herald Square) to 7th Ave.*

Plymouth Rock, the Mayflower II, and the Plimouth Plantation provide a close-up look into the Pilgrims history. Google Earth can provide a broader view of their experience. The pilgrims set out across the ocean in the Mayflower from Plymouth, England in September 1620. Cape Cod, Massachusetts was the first land the Pilgrims saw when they arrived in America in November 1620. Zoom out to get a regional view of Cape Cod – it is a huge moraine left behind by the Laurentide Ice Sheet during the last ice age. Check the ocean depth (elev in the Status Bar) south of Cape Cod around Nantucket and Martha's Vineyard. The ocean is very shallow here due to outwash as the glaciers retreated. Notice the many kettle hole ponds on the cape created when giant chunks of ice left behind by the glaciers melted (41°56'22"N 70° 0'13"W for example). The Pilgrims were looking for the mouth of the Hudson (New York) and first headed southeast from the Cape, but the shallow waters around Nantucket prevented them from continuing past Nantucket. So they headed back up around the cape and anchored near present-day Provincetown, Massachusetts, where they signed the Mayflower Compact. They explored the coast for some time and eventually settled in present-day Plymouth in December 1620. Explore the coastline yourself to see where you might have landed.

Christmas: *Bethlehem, Israel or the North Pole. Do you think Santa lives at the Geographic North Pole (90°N 0°W or 90°N 0°E) or the Magnetic North Pole (84°58'N and 132°21'W in 2010)?*

New Year's Eve: *Times Square, New York City, New York*

Playing Sports

Bird's Nest Stadium, Beijing, China. (Google Earth Image 2012 DigitalGlobe, 39°59'30.28"N 116°23'28.91"E, eye alt 2500 ft)

The list of famous sports venues is endless, but here are a few to get you thinking:

Ṯhe Olympics
Olympia, Greece (37°38'21"N 21°38'00"E) – *For a map of the original buildings: http://en.wikipedia.org/wiki/File:Plan_Olympia_sanctuary-en.svg*
Wembley Stadium, London, England – *Used for soccer=football in the 2012 Olympics and for most events in the 1948 Olympics – the first Olympics after WWII.*
Beijing National Stadium, Beijing, China – *Also known as the Birds Nest; use the Clock to watch it under construction.*

Baseball
In addition to the Major League Baseball stadiums or your favorite Little League field, take a look at these famous places:
Howard J. Lamade Stadium, South Williamsport, Pennsylvania (41°13'49"N 76°58'50"W) – *Little League World Series*
Baseball Hall of Fame, Cooperstown, New York – *Look for Doubleday Field and Otsego Lake, known to James Fenimore Cooper as Glimmerglass and prominent in* The Last of the Mohicans

Football
Rose Bowl Stadium, Pasadena, California – *1/12/2004 on the Clock slider shows the field decorated for the 2004 Rose Bowl Game.*
Reliant Astrodome, Houston, Texas – *The world's first domed stadium.*
Golf
Augusta National Golf Club, Augusta, Georgia – *Notice how bright the sand traps are against the uniformly green grass.*
The Royal & Ancient Golf Club Of St Andrews, Fife, Scotland (zoom out to see the course to the north along the coast) – *Golf has been played here since 1400 AD. Use the Clock to see what the course looked like in 1945. Open the Scotland 1790 map in the [Rumsey Maps] layer to see that the course was noted (although the maps are not well co-registered).*

Biking
Tour de France – *The course changes every year but always ends in Paris. Search the Internet for a kmz file and download it. I found one for 2010 at: http://www.gearthblog.com/blog/archives/2010/07/2010_tour_de_france_in_google_earth.html. Also look for kmz files for Giro D'Italia in Italy and Vuelta a España in Spain, which, together with the Tour de France, make up the grand tours.*

Dog Sled Racing
Iditarod from Anchorage to Nome, Alaska – *Search the Internet for a kmz file and download it. I found one at http://www.gearthblog.com/blog/archives/2007/03/dog_sled_races_in_go.html*

Yukon Quest between Fairbanks, Alaska and Whitehorse, Yukon – *The course follows the route of the 1890s Klondike Gold Rush between Fairbanks, Alaska, Dawson City, Canada, and Whitehorse, Canada (remember <u>Call of the Wild</u>?).*

Soccer = Football
Soccer City Stadium, Johannesburg, South Africa – *2010 World Cup; fun to watch it being built using the Clock.*
Estadio Azteca in Mexico City, Mexico – *The only stadium to have hosted two World Cups.*

Swimming
The English Channel – *The best way across the English Channel is via the Strait of Dover, the narrowest part of the English Channel, from Admiralty Pier in Dover, England to Gris Nez, France. Search for Gris Nez. Zoom in and use the Look joystick to see if you can see the White Cliffs of Dover, which are made of chalk, a sedimentary rock.*
Hawaii Ironman, Hawaii (the Big Island) – *The swim start and finish and bike start is on Kailua Pier in Kona (19°38'21"N 155°59'49"W); the bike portion heads north to Upolu Point, then heads back south to Kona Surf Hotel where the run starts. The run heads north to Keahole Point and the back to the Kailua Pier. Zoom in to find the roads and the black lava flows across which the roads travel.*

Horse Racing:
Triple Crown – *The Kentucky Derby is the first at Churchill Downs, Louisville, Kentucky. The Preakness Stakes at Pimlico Race Course, Baltimore, Maryland follows, and finally, the Belmont Stakes at Belmont Park Race Track, Elmont, New York completes the Triple Crown. Which track is the longest? Do you recall how to calculate the circumference of an oval given the lengths of its two axes, L and W? (hint: $\pi W + 2(L-W)$)*
Grand National Steeplechase at Aintree Racecourse, Liverpool, England – *The course is also used for motor racing. Zoom in to follow the horses' jumps around the 4.5 mile course.*

Car Racing
Daytona International Speedway, Daytona Beach, Florida – *Daytona 500 NASCAR race*
Indianapolis Motor Speedway, Indianapolis, Indiana – *Indianapolis 500-Mile Race and Brickyard 400*

Cricket
Lords Cricket Ground, London, England – *Use the Clock to find the checker board mowing pattern. Was the field in use in 1945?*
Melbourne Cricket Ground, Melbourne, Australia – *Use the Clock to find a variety of colors and patters on the field due to construction, mowing patterns and field usages.*

Basketball
Madison Square Garden, New York City, New York – *The Garden sits on top of Penn Station – look to the northwest to see the tracks.*
Rucker Park, Manhattan, New York (40°49'46.65"N 73°56'11.22"W) – *A small basketball court in Harlem across the East River from Yankee Stadium. Many NBA players started on this court.*

Tennis
Note the different colors of the courts that are grass, clay and hard court:
Melbourne Park, Melbourne, Australia – *Australian Open*
Stade Roland Garros, Paris, France – *French Open*
All England Club, Wimbledon, London, England – *Wimbledon*
USTA Billie Jean King National Tennis Center, Queens, New York City, New York – *US Open; look for Arthur Ashe Stadium*

Marathons
Badwater Ultramarathon – *Death Valley to Mt. Whitney. Draw a Path between Death Valley (Badwater Basin is the lowest point) and Mt. Whitney and highlight it in the Places Panel. Look in the Google Earth Menu Bar under "Edit" for "Show elevation Profile" and select it. An elevation profile will appear on the screen to show you how many ups and downs are associated with this grueling run.*
Boston, New York and London Marathons – *Look on the Internet for kmz or kml files of these famous marathons.*

Bull Fighting
Plaza de Toros de Las Ventas, Madrid, Spain
La Maestranza, Seville, Spain

Running of the Bulls
Cuesta Santo Domingo, Pamplona (42°49'10.47"N 1°38'45.45"W) – *Beginning here, turn on the [Roads] layer and follow these streets:*
Calle de los Mercaderes (42°49'5.73"N 1°38'37.77"W)
Calle de la Estafeta (42°49'5.89"N 1°38'34.13"W)
Plaza del los Toros (42°48'57.76"N 1°38'22.02"W)

Making Movies

Field of Dreams movie setting in Dyersville, Iowa. (Google Earth Image USDA Farm Service Agency, 42°29'49.92"N 91°03'15.20"W, eye alt 4600 ft)

Movie scouts have the unique task of finding the settings for movies that help tell the story and maintain the integrity of the setting laid out by the author or screenwriter. Sometimes the settings are in the actual location where the story takes place, and sometimes the settings are buried in the movie studios or the back lots of Hollywood. You can start your exploration of movie settings in California with the major movie studios, although most of the actual sets are indoors.
Walt Disney Studios, Burbank, California – *Look for the Sorcerer's Apprentice hat at 34°9'18.99"N 118°19'24.67"W. Use Street View to see it.*
Warner Brothers Studios, Burbank, California
Universal Studios Hollywood, Los Angeles, California – *The theme park is at the center of your search and the studios where filming actually happens are the buildings with white roofs to the northwest of the theme park.*

The filming of Harry Potter required a variety of settings, many of which are scattered all over England and Ireland. Here are a few – perhaps you know of more:
Dursley house (51°24'31.08"N 0°43'17.48"W) – *The house at #4 Privet Drive is actually 12 Picket Post Close, Berkshire, England.*
Kings Cross Station – *A real train station in London.*
Glenfinian Viaduct, Scotland (56°52'34"N 5°25'52"W, eye alt 2000 ft) – *The Hogwarts Express passes over this viaduct on its way to Hogwarts. Use the [3D Buildings] layer to find it.*
Alnwick Castle, Alnwick, England – *This is one of the castles used to represent Hogwarts.*

The Sound of Music features one of the most beautiful movie settings centered in Salzburg, Austria.
Nonnberg Abbey, Salzburg, Austria – *Maria's abbey.*
Schloss Leopoldskron, Salzburg, Austria – *The von Trapp house on the water.*
Mondsee Cathedral, Mondsee, Austria (47°51'22"N 13°21'4"E) – *The cathedral where Maria was married.*
And of course the green hills all around Salzburg

Hitchcock films include many memorable film locations. I wonder if Alfred Hitchcock is somewhere in a Google Earth image!
The Birds – *Bodega Bay, California*
North by Northwest – *Mount Rushmore, South Dakota*
The Man Who Knew Too Much – *Albert Hall, London, England*
Psycho – *The Psycho House is in the Universal Studios lot (34°8'10.49"N 118°20'48.58"W, eye alt 1000 ft); the Bates Motel is at 34°8'11.89"N 118°20'47.97"W. Notice the crashed airplane parts behind the Psycho House from some other movie.*

Other movies with memorable settings:
Close Encounters of the Third Kind – *Devil's Tower National Monument, Wyoming at 44°35'25"N 104°42'54"W.*
Field of Dreams – *Lansing Road, Dyersville, Iowa at 42°29'52N 91°3'18"W.*
Jaws – *Martha's Vineyard, Massachusetts, mostly in the town of Menemsha. This island was chosen because the shallow sandy ocean helped with the shark scenes. Zoom out to an eye alt of 40 mi to see how extensive the shallow waters are surrounding the island.*
Rocky – *The Philadelphia Museum of Art, Philadelphia, Pennsylvania*
Star Wars – *The village from Episode I lies at 33°59'39"N 7°50'34"E. The canyon where R2-D2 was attacked by the Jawas is called Sidi Bouhlel – its mouth is at 34°2'1"N 8°16'54"E. Turn on the [Photos] layer at an eye alt of 2500 ft and look for the photos that follow the canyon. It is also the site where Indiana Jones threatens to blow up the Ark in* Raiders of the Lost Ark.

Thumbing Through National Geographic Magazine

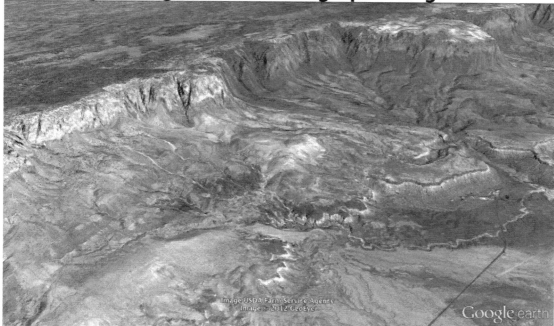

The Vermillion Cliffs in the Grand Canyon as seen using the Look joystick. (Google Earth Image USDA Farm Service Agency, 2012 GeoEye, 2012 TerraMetrics, 36°44'33.32"N 111°46'46.20"W, eye alt 12000 ft)

How many National Geographic Magazines do you have in your basement? Now there is a new use – drag them out, page though them until you find an interesting place, and search for it on Google Earth. Here are some places I discovered using the February 2012 issue with the Wegman Weimaraner on the cover. Not much to say with Google Earth about dogs except that I was in Mozambique last year and all the dogs looked like the one on page 52.

Maarjah, Afghanistan – *Search for the region shown on page 12 and zoom out to appreciate the "drought-pocked" earth where these sheep live. On page 14, a very different part of Afghanistan is presented – one that has several sources of water. Search for Rushan, Afghanistan and compare the two regions.*

Rahmatullah Mosque, Sumatra (5°29'39.38" N 95°14'07.11"E) – *Page 72 and 73 show a photo of this mosque soon after the December 2004 tsunami. This one was a little hard to find because Lampuuk, Sumatra does not show up on Google Earth. But by googling Rahmatullah Mosque on the Internet, its lat/long was discovered. The mosque is apparent along with some very nice houses surrounding it. But use the Clock to go back to 2005 to see the devastation of the tsunami.*

Astana, Kazakhstan – *The buildings in Astana are a bit difficult to find, but start with "Presidential Palace, Astana, Kazakhstan" and the map on page 89, and all will be clear. This is a great place to look at the [3D Buildings] layer – especially the Khan Shatyr shopping mall.*

Vermillion Cliffs – *A trip to the Grand Canyon is always a must in Google Earth. In the article on page 110, the Vermillion Cliffs are the focus. Although "Vermillion Cliffs" does not get you to the right place using the Search Panel, you can find these colorful cliffs by searching for the Navajo Bridge, adding the [Roads] and [Borders and Labels] layers, and zooming to an eye alt of about 40 mi. Then use the map on page 119 to orient yourself and find the cliffs, and zoom in to about 30,000 ft. Use the Move joystick to travel along the cliffs, which gives you an amazing 3D perspective. Use the Look joystick to get a better view of the sandstone layers.*

In the last article, photographers follow a group of nomadic people in Meakambut, Papua New Guinea (yes, you really can find this place by searching for "Meakambut, Papua New Guinea"!). Zoom in and use the Move joystick to move around to get an idea of the dense forested landscape in which these people live, and how hard it would be to find a hospital if your child were sick.

Launching the Manned Space Program

Launch Pad 39A at the Kennedy Space Center in Florida where the Apollo's and many Shuttles launched. The Shuttle is on the launch pad in this image (image date 2/4/10) as can be see by looking for the orange external Fuel Tank and the two solid Rocket Boosters. The Shuttle itself is hidden beneath the swing arm use to provide access to the Shuttle as it is prepared for flight. (Google Earth Image, 28°36'29.52"N 80°36'14.92"W, eye alt 4000 ft)

The manned space program began when the Soviet Union launched Yuri Gagarin from the Baikonur Cosmodrome launch facility (Pad 1/5 at 45°55'13"N 63°20'31"E) in Kazakhstan (then part of the Soviet Union). The United States soon followed with the launch of Al Shepard in a Mercury rocket from Kennedy Space Center (KSC) in Florida (Redstone Launch Complex 5, 28°26'22"N 80°34'24"W). Zoom out over KSC to see that one reason the east coast of

Florida was selected for a launch site was that any failed rockets, which always launch heading east, would land in the ocean.

The space programs in both countries evolved first through competition, and now through cooperation. Today the International Space Station is usually occupied by at least one Russian and one American. The station was built using the Space Shuttle, which was launched from KSC using solid rockets and a liquid fuel tank built at the Michoud Assembly Facility, New Orleans, Louisiana. The external tanks were transported by barge to KSC. You can see the Shuttle (or at least the solid rockets and the external fuel tank) on Launch Pad 39A (28°36'31"N 80°36'15"W) on 4/20/2012 and 1/2320/09. This launch complex was decommissioned in 2007 and the remainder of the Shuttle launches were from Launch Pad 39B. Also look for the landing strip and the Vehicle Assembly Building, where the Shuttles are integrated with the solid rockets and external fuel tanks. The Shuttle usually landed back at KSC, but sometimes landed at Edwards Air Force Base in California when the weather in Florida did not cooperate. Look for the runways at Edwards, and the giant compass rose etched on the dry lakebed (34°57'16"N 117°52'24"W). Soyuz capsules are launched on Soyuz rockets from the Baikonur Cosmodrome, Kazakhstan. They land in the steppe of Kazakhstan (look between the cities of Arkalyk, Karaganda, and Zhezkazgan in Kazakhstan – quite a wide area!). All of NASA's manned missions are controlled from Johnson Space Center in Houston, Texas, and all of Russia's manned missions are controlled from Mission Control Center Moscow in Kordyov, Russia (55°54'45"N 37°48'37"E).

To see all of the launch sites, including those for unmanned missions, go to http://www.spacelaunchreport.com/padsites.html.

Flying to the Moon

Google Earth view of the Apollo 11 Lander on the Moon. The Lander is a 3D Building but the lunar surface is a real image. (Google Earth Image NASA/USGS, 0°40'53.55"N 23°27'34.84"E)

Every year I give a talk to some third grade classes about Apollo and the Moon. When I introduce the astronauts, I tell the students that these men were selected because they were smart, physically fit, good team players, and cool-headed. Then I ask the students what 'cool headed' means, and I get some interesting answers, including that the astronauts are "bald", "forgot their helmet", or "had to shave for the trip".

You have been exploring Earth with Google Earth, but you can also explore our Moon using the Planet tool. Click to the Moon and zoom in so the Moon fills the 3D Viewer. Zoom to an eye alt of about 200 mi and compare the number of craters on the dark lava to the number on the brighter areas of the Moon. The more craters, the older the surface, so the lava seas are a relatively new surface on the Moon. Search for Slipher – a crater on the far side of the Moon. Zoom to an eye alt of about 60 mi. Slipher Crater is about 45 miles in diameter. Use the Ruler and click on the [Place Names] layer to make sure you are looking at the correct crater. Find a smaller crater on the west rim of Slipher Crater. The overlapping crater is younger than Slipher, the one being overlapped. This is one way scientists can age date craters on the Moon.

Six Apollo missions landed on the Moon. Each mission had three astronauts, two of whom walked on the Moon, and one who stayed in the Command Module. When the astronauts were done with the equipment they needed on the Moon, they left it on the surface to make room in their spaceship for Moon rocks. There are 24 moon boots on the Moon, but they are not all in one place because each Apollo mission landed in a different region. Look in the Layers

Panel for [Moon Gallery] and open it. Click on the [Apollo Missions] layer to see a view of the Moon that shows all six landing sites. Note that all the landing sites are on one side of the Moon – the side of the Moon that always faces us so the astronauts could be in constant communication with Mission Control on Earth.

Open and click on the [Apollo 11] layer and zoom in to find the Lander, which is a 3D model. Find the camera icon labeled "The View from South-East of the Lander Module" and click on it. Then click on the picture in the popup window to move to a panorama of the landing site taken by the astronauts. Use the navigation window in the upper right of the 3D Viewer to move around the photo. Look for the Lander, the shadow of the astronaut taking the picture (usually it was Neil Armstrong), footprints (which would still be there today since there is no water or wind to wash them away), the other astronaut (Buzz Aldrin if Armstrong is holding the camera), and small craters.

Landing on Mars

Spirit's view near the end of its journey. (Google Earth Image, 14°36'10.82"S 175°31'33.69"E)

My daughter's Girl Scout troop was touring the Jet Propulsion Lab and had the opportunity to visit the Mars Yard where rovers were tested. At that time, the Mars Yard was an area in a large room that was filled with sand – like a giant sandbox. The girls were invited onto the sand to get a closer look at the rovers. A parent who had joined us on the tour asked, "Is that sand from Mars?" An innocent question, but if we had ever brought back any sand from Mars, it would be in a vault, and would surely not fill a room, and would definitely not be something Girl Scouts could walk on.

Anyone can, however, explore Mars without touching its sand using Google Earth. Click on the Planet tool and select "Mars". Zoom to an eye alt of about 1000 mi. Use the Move joystick to fly around the planet. You may recognize craters, volcanoes and valleys. Notice that the color of Mars is reddish-orange all over the planet due to high concentrations of iron oxide in the soil. Use the Search Panel to search for Valles Marineris and Olympus Mons. As with Earth, the Ruler and the elev can help you determine the scale of these prominent Mars features.

Zoom out to a global view of Mars and look in the Layers Panel for [Global Maps]. Choose [Colorized Terrain] and Mars will all of a sudden become very colorful. The topography of Mars has been used to color each pixel on the globe according to its elevation. The result is a false-color picture that emphasizes its terrain. Check the scale in the lower corner. Now it is easy to find Olympus Mons and Valles Marineris without using the Search Panel!

In some areas, high-resolution imagery taken from Mars orbiters can be viewed. Click off the [Colorized Terrain] layer and click open [Featured Satellite Images] in the Layers Panel. Look in the 3D Viewer for the red and yellow squares. The red squares indicate imagery from the High Resolution Imaging Science Experiment (HiRISE) on the Mars Reconnaissance Orbiter (MRO); zoom in over these squares until you see the grey rectangles. Zoom further into the rectangles to see a detailed view of the surface features. You can click on the red boxes to learn more about these features. The yellow boxes mark imagery from the Context Camera (CTX), also on MRO. Click on a yellow box and then on "Load this image" so see these areas. Additional high-resolution images can be found under [Spacecraft Imagery] in the Layers Panel. Here are some favorite regions with high-resolution imagery – use the Search Panel to find them.

Proctor Crater
7°43'33"N 6°48'13"W
Northwest edge of Schiaparelli Crater

Recall that the rovers Spirit and Opportunity landed on Mars in 2004. Spirit lasted until 2009 and Opportunity is still working as of 2015. Click open the [MER Spirit Rover] layer in the [Mars Gallery] → [Rovers and Landers] and notice the path of the rover and the cameras that mark the locations of panorama photographs. Look for the panorama titled "Legacy Panorama on Spirit's Way to 'Bonneville'" and click on it. When the popup window appears, click on "Fly into this high-resolution photo". Use the controls in the upper right corner of the 3D Viewer to navigate around the panorama. Do you see part of the rover and its tracks in the sand? You can also follow the latest rover, Curiosity, as it explores Gale Crater looking for evidence that life could have existed on Mars. Click on the [MSL Curiosity Rover] layer and zoom in to investigate the [Targets of Interest] under the [Gale crater landing site]. Where would you take this rover?

Google Earth Books for Teachers

Association of Educational Publishers Distinguished Achievement Awards, Winner, 2013

If you are a teacher, take a look at these Google Earth books designed for grades 1-2, 3-5, and 6-8. The 6-8 book is also appropriate for high school teachers. The books include highly engaging lessons, and easy-to-use, step-by-step instructions that integrate this Google Earth sites into social studies, science, mathematics, and English language arts curriculum. The Using Google Earth books show teachers how to help their students start their own .kmz folders and fill them with layers of locations that connect their own lives to their curriculum, and to build cross-curricular connections.

Holt, JoBea, 2012. Using Google Earth: Bring the World into Your Classroom, Level 1-2, Shell Education Pub, 256pp.

Holt, JoBea, 2012. Using Google Earth: Bring the World into Your Classroom, Level 3-5, Shell Education Pub, 256pp.

Holt, JoBea, 2012. Using Google Earth: Bring the World into Your Classroom, Level 6-8, Shell Education Pub, 256pp.

Made in United States
Orlando, FL
27 June 2022